E-MRS Fall Symposium I: Solutions for Critical Raw Materials Under Extreme Conditions (E-MRS 2017)

IOP Conference Series: Materials Science and Engineering
Volume 329

Warsaw, Poland
18 - 21 September 2017

Editors:

Maria Letizia Ruello
Daniele Valerini
Saurav Goel
Paivi Kivikyto-Reponen

ISBN: 978-1-5108-6039-1
ISSN: 1757-8981

Printed from e-media with permission by:

Curran Associates, Inc.
57 Morehouse Lane
Red Hook, NY 12571

Some format issues inherent in the e-media version may also appear in this print version.

Printed by Curran Associates, Inc. (2018)

For permission requests, please contact the Institute of Physics at the address below.

Institute of Physics
Dirac House, Temple Back
Bristol BS1 6BE UK

Phone: 44 1 17 929 7481
Fax: 44 1 17 920 0979

techtracking@iop.org

Additional copies of this publication are available from:

Curran Associates, Inc.
57 Morehouse Lane
Red Hook, NY 12571 USA
Phone: 845-758-0400
Fax: 845-758-2633
Email: curran@proceedings.com
Web: www.proceedings.com

Table of contents

Volume 329

E-MRS Fall Symposium I: Solutions for Critical Raw Materials Under Extreme Conditions

18–21 September 2017, Warsaw, Poland

Accepted papers received: 5 March 2018
Published online: 26 March 2018

Preface

Papers

Preface

Maria Letizia Ruello
Università Politecnica delle Marche, Brecce Bianche, 60131 Ancona, Italy
E-mail: m.l.ruello@univpm.it
Daniele Valerini
ENEA - Italian National Agency for New Technologies, Energy and Sustainable Economic Development, S.S. 7 Appia - km 706, 72100 Brindisi, Italy
E-mail: daniele.valerini@enea.it
Saurav Goel
School of Aerospace, Transport and Manufacturing, Cranfield University, Building No. 90, Bedfordshire, MK430AL, UK
E-mail: s.goel@cranfield.ac.uk
Päivi Kivikytö-Reponen
VTT Technical Research Centre of Finland Ltd, Sinitaival 6, 33101 Tampere, Finland
E-mail: paivi.kivikyto-reponen@vtt.fi

The Symposium "Solutions for Critical Raw Materials Under Extreme Conditions" was held at Warsaw University of Technology, Warsaw, Poland, from September 19th to 21st, 2017 in the framework of the eminent E-MRS Fall Meeting. The meeting addressed the difficulties in the access to critical raw materials (CRMs) that are expected to depress industrial sectors vital especially for our society across Europe.

Global and local crises have clearly shown that life in a modern society ultimately depends upon readily available and affordable material supply. In this scenario, a clear goal is to reduce the dependence of industrial sectors on critical raw materials almost entirely imported from foreign countries, often with extremely volatile prices because of geopolitical reasons.

In particular, the Symposium was mainly focused to a well-defined sector concerning the materials meant to be used under extreme conditions of temperature, loading, corrosion, friction, and wear. In this framework, it was aimed to find solutions in related technological fields, such as manufacturing, machining, transportation, energy and construction. .

The Symposium provided a unique platform for a wide range of researchers from various international locations who are engaged in performing cutting-edge multidisciplinary research on strategic CRMs substitution, new material design, LCA, recycling, circular thinking, that is expanding and consolidating around a few initial catalytic nuclei of growth. The papers collected in the proceedings reflect this wide approach to the challenge.

This Symposium was organised jointly by two of the largest networks currently aiming to find sustainable solutions for materials in extreme conditions: the COST Action CRM-EXTREME (http://www.cost.eu/COST_Actions/ca/CA15102) and the network of infrastructure EXTREME under the KIC EIT RawMaterials (http://www.network-extreme.eu/), bringing together a large number of leading researchers from within and outside Europe.

IOP Conf. Series: Materials Science and Engineering **329** (2018) 011001 doi:10.1088/1757-899X/329/1/011001

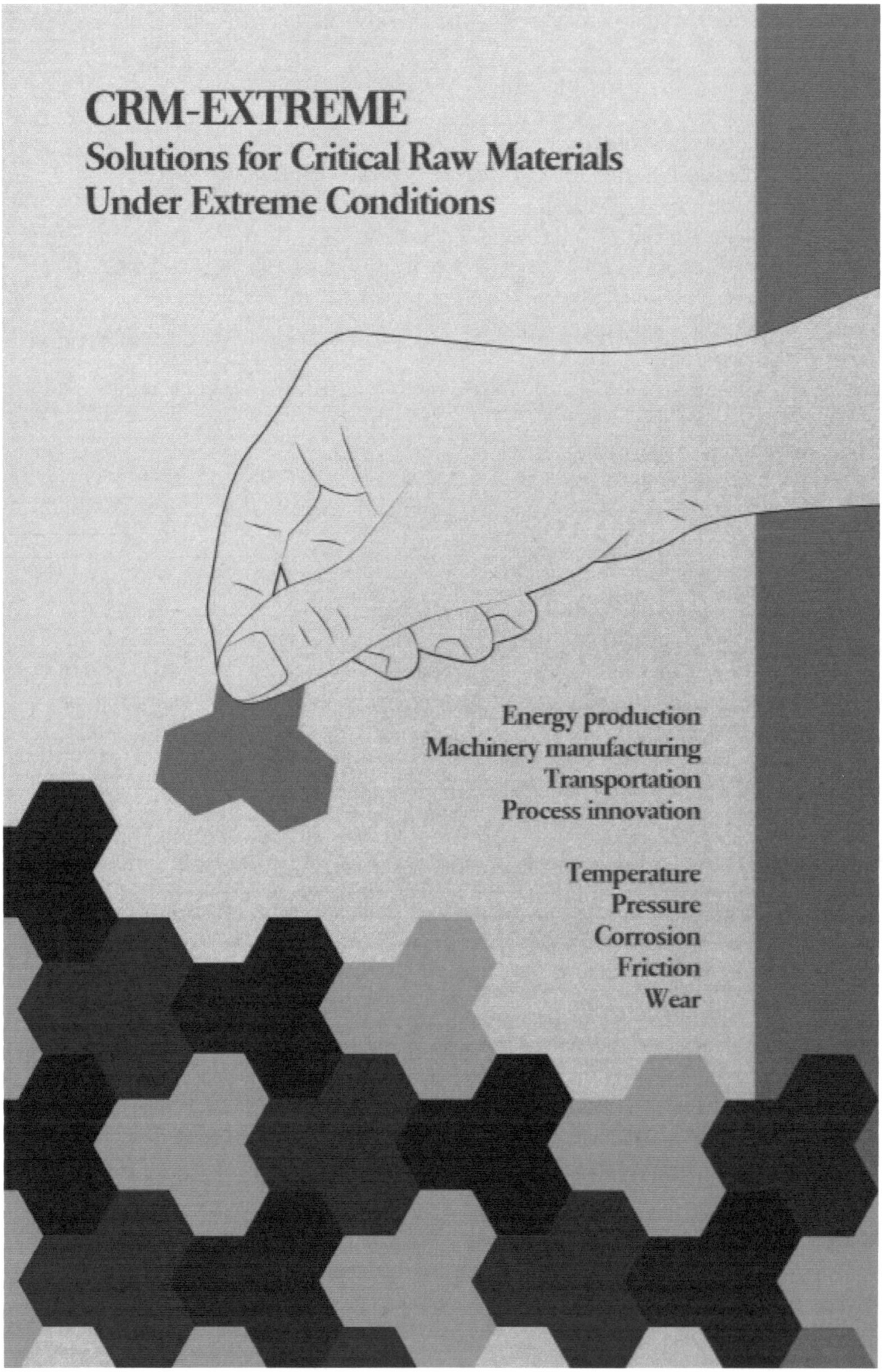

Peer review statement

All papers published in this volume of *IOP Conference Series: Materials Science and Engineering* have been peer reviewed through processes administered by the proceedings Editors. Reviews were conducted by expert referees to the professional and scientific standards expected of a proceedings journal published by IOP Publishing.

VERAM, for a sustainable and competitive future for EU Raw Materials

A Mobili[1], F Tittarelli[1,2], G M Revel[3] and P Wall[4]

[1] Department of Materials, Environmental Sciences and Urban Planning, Università Politecnica delle Marche, Research Unit INSTM, Ancona, Italy
[2] Institute of Atmospheric Sciences and Climate, National Research Council (ISAC-CNR), Bologna, Italy
[3] Department of Industrial Engineering and Mathematical Sciences, Università Politecnica delle Marche, Ancona, Italy
[4] European Technology Platform on Sustainable Mineral Resources c/o EuroGeoSurveys, Brussels, Belgium

E-mail: f.tittarelli@univpm.it

Abstract. The project, VERAM "Vision and Roadmap for European Raw Materials", aims to deliver a mapping of on-going initiatives on non-food, non-energy raw materials (including metals, industrial minerals, aggregates and wood) at European, Member State, and regional levels both from the Research and Innovation (R&I), industry, and policy perspectives. Moreover, based on a comprehensive gap analysis, VERAM will propose a common long term 2050 Vision and Roadmap in coordination and cooperation with all stakeholders across the value chain. For the first time, two European Technology Platforms (ETPs) together with their corresponding European Research Area Networks (ERA-NETs) are joining forces to develop a common roadmap.

1. The project

The Vision and Roadmap for European Raw Materials (VERAM) proposal answers to the Horizon 2020 call H2020-SC5-13d-2015, which looks for raw materials Research and Innovation (R&I) coordination.

VERAM aims to provide the umbrella and coordination function for raw materials related R&I activities that address security and sustainability of supply from EU and non-EU sources and creates new synergistic effects due to linking of biotic and abiotic resource sectors. It intends to develop strategies to bring innovations to market and develop recommendations for future research and innovation related policy actions.

It is for the first time that two European Technology Platforms (ETPs) together with their corresponding European Research Area Networks (ERA-NETs) join forces to develop a common roadmap. The VERAM partnership is in fact constituted by the strongest possible representation of raw material sectors in Europe. It unites five European Technology Platforms (ETPs) and two European Research Area Networks (ERA-NETs), the newly established EIT Knowledge and Innovation Community (KIC) on Raw Materials. Since several of the partners of VERAM lack legal entity, they have chosen one of their members to act as the legal representative; this is the case of the two ERA-NETs and three of the five ETPs.

The ETPs have been established during the sixth Framework Programme as industry-led stakeholder fora to develop short to long-term strategic research and innovation agendas and roadmaps for actions at EU and national level.

The research agendas [1-5] and roadmaps produced by the ETPs and the relevant ERA-NETs form the basis for the so-called Strategic Implementation Plan (SIP) of the European Innovation Partnership (EIP) on RM. The SIP identifies actions to achieve the objectives of the EIP RM which includes improved coordination of research and innovation activities. It makes sense to involve the ETPs and ERA-NETs in an action to better coordinate the research and innovation activities as well as to evaluate the different ETP research agendas and roadmaps to identify synergies and gaps.

1.1. The project Consortium

The five ETPs in VERAM Consortium are: the European Technology Platform on Sustainable Mineral Resources (ETP SMR), which is also the project leader; the Forest-based Sector Technology Platform (FTP); the European Construction Technology Platform (ECTP), represented by Università Politecnica delle Marche (UNIVPM); the European Technology Platform for Sustainable Chemistry (SusChem), represented by the European Chemical Industry Council (Cefic) and the European Technology Platform on Advanced Materials (EuMaT), represented by the Vlaamse Instelling voor Technologisch Onderzoek (VITO).

The two ERA-NETs are: the Network on the Industrial Handling of Raw Materials for European Industries (ERA-MIN), represented by the National Centre for Scientific Research (CNRS), and the ERA-NET on wood material science and engineering (WoodWisdom.NET), represented by Fachagentur Nachwachsende Rohstoffe e.V. (FNR).

In addition, the VERAM consortium comprises RINA Consulting S.p.A. (RINA), Forschungszentrum Juelich GmbH (FZ Juelich), Tecnalia Research & Innovation (Tecnalia) and VTT Technical Research Centre of Finland Ltd. (VTT).

In VERAM, the biotic raw materials value chain is represented by both FTP and WoodWisdom.NET. In particular, FTP has 23 National Support Groups and deals with inventory, harvesting, processing, recycling and substitution of biotic raw materials. The forest-based sector in Europe comprises a wide variety of products and services, ranging from paper, packaging, tissue paper, furniture, carpentry and construction materials made from solid wood and wood-based panels to textile fibers, biofuels and special chemicals. Healthy and resilient forests are a prerequisite for a sustainable supply of raw materials as well as for other goods and services provided by forest ecosystems (including carbon sequestration, clean water, erosion protection, biodiversity, aesthetic landscapes and recreation).

The abiotic raw materials value chain, which comprises metals, minerals and aggregates, is represented by both ETP SMR and ERA-MIN. The European Technology Platform on Sustainable Mineral Resources (ETP SMR) unites stakeholders from industry, the research community, public authorities, the financial community, regulators, consumers and civil society to jointly develop a common vision. In particular, the European minerals industries involved include metal ores, industrial minerals, ornamental stones, aggregates, smelters as well as technology suppliers and engineering companies. The ETP SMR deals with exploration, extraction, processing, recycling and substitution of inorganic raw materials.

1.2. Objectives

Research and innovation activities in the area of primary and secondary raw materials - their use, reuse, and substitution - are dispersed between different players and supported by a variety of fragmented funding sources both on the EU and on Member States levels. The current situation is not optimal for unleashing the full innovation potential of the concerned value-chains. VERAM aims to create a critical mass by bringing together the stakeholders represented by the ETPs, the ERA-NETs, as well as the EIT KIC on Raw Materials.

The multi-disciplinary nature of research related to raw materials requires a coordinated multi-disciplinary approach on all levels. However, only around 15% of European publicly funded research is coordinated jointly in the EU's Research Framework Programme. The remaining 85% is funded via national or regional programmes. These funding programmes are very important but too often they unnecessarily lead to duplication of research efforts and they often lack the scope and depth required of the EU programmes. Incorporating as a part of the VERAM proposal's partnership the two most central ERA-NET partnerships in the area of raw materials will help to mobilize and orchestrate the full 100% of European public research and innovation funding for the implementation of the 2050 roadmap envisioned.

The VERAM project aims to envelope all relevant aspects of non-food, non-energy raw materials related research and innovation, including metals, industrial minerals, aggregates and wood. It challenges the current compartmentalization into scientific disciplines and fragmentation into industry sectors to increase synergies and facilitate uptake of research results. In doing this, VERAM identifies the common interests, respecting also the individual visions of each stakeholder group in recognition of the specific conditions that they have to comply to.

It will provide an innovation reference point for the EIT Raw Materials (former KIC Raw MatTERS) and will coordinate the network of people involved in the EIP RM Commitments and in the proposals funded under Horizon 2020.

It will provide a web-based platform for identifying gaps and complementarities and bridge these.

VERAM will also advise the European Commission and Member States on future research needs and policies to stimulate innovation and assist in overcoming the fragmentation in the implementation of the EIP RM Strategic Implementation Plan.

During its duration, VERAM will look for **capacity building**, **transfer of knowledge**, mutually beneficial **information exchange**. It will also **encourage cross-fertilization** between actions undertaken by different raw material industries and will **speed-up exploitation of breakthrough innovations**.

As one of the main deliverables of the proposal, a **common long term 2050 Vision and Roadmap for the relevant raw materials** will be presented. The vision and roadmap will also show the path to achieve the European Commission's ambitious target of 80% reduction in CO_2 emissions by 2050.

In addition to the contributions to the requirements of the call, VERAM promotes innovation in the following ways:

- **Transparency**. A significant improvement in the exchange and use of the information available in the projects between the different stakeholders will be performed by mapping of ongoing research activities and research funding.
- **Visibility**. A higher degree of visibility to society, industry and policy makers for the innovation and research need.
- **Coordination**. Less risk of wasting resources on overlapping tasks and activities.
- **Complementarities**. Combining complementary research results and bridging detrimental technology-gaps by sharing of knowledge and cross-fertilization.
- **Critical mass**. A broader stakeholder base and critical mass for exploitation activities by bringing together ETPs and ERA-NETs relevant for the EIP-RM challenges.
- **Inclusion**. Involvement of the citizens, not only experts and industrial operators touched by the innovation in the field of Raw Materials, but extending the level of knowledge to the whole European people performed by stakeholder consultations.
- **Knowledge**. Easier access to information about national and international research and innovation projects through a web-based research and innovation portal.
- **Exploitation**. The funding-related "valley of death" for identified key innovation areas will be bridged by the incorporation of regional and national funding providers, as well as the support of the European Commission, in this way new business models and higher participation in innovation activities are expected.
- **Cooperation**. Improved mutual understanding and trust will ease the forming of future project consortia.

A successful execution of the VERAM proposal will help the society to reap the benefits of a growing circular economy and to address the Societal Challenges by a more sustainable and resource-efficient use of raw materials.

2. The roadmapping process

The roadmapping process is a flexible technique that is used to support strategic and long-range planning. The approach provides a structured (and often graphical) means for exploring and communicating the relationships between evolving and developing markets, products and technologies over time [6]. The roadmapping process provides a way to identify, evaluate, and select strategic alternatives that can be used to achieve desired science and technology objectives [7].

Broadly, two cultures of roadmapping can be identified. The first is the technology roadmapping, in which roadmapping is approached as a normative instrument to identify relevant technologies and align them with explicit product plans and related action steps. In this culture, the roadmapping process is a systematic practice aimed at product development. The second is the emerging culture of strategy roadmapping, in which the roadmapping is perceived more as a dynamic and iterative process that produces weighed crystallizations of long-term vision, and short- to medium-term strategies to realize this vision. There, the roadmaps can be seen as knowledge umbrellas that depict a large-scale strategy picture of a system [8]. In this regard, the VERAM roadmapping process belongs to the second culture of roadmapping [9], which its graphical means is schematically given in Figure 1.

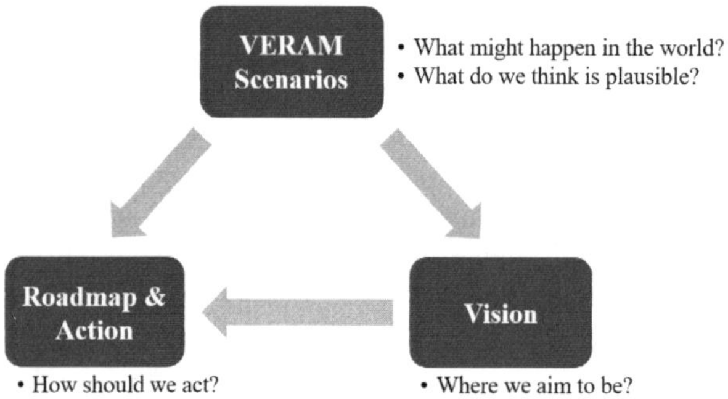

Figure 1. A model of an ideal Roadmapping process for VERAM.

3. The (preliminary) results

3.1. A common terminology

As indicated in the Strategic Implementation Plan for the EIP RM, terminology and reporting standards related to raw materials are heterogeneous throughout the EU member states [10]. Moreover, different areas within the raw materials sectors use different terminologies and classifications for raw materials and their properties. Since VERAM's first objective is to facilitate information exchange, overcome fragmentation and promote cross-fertilization, a common terminology related to definitions of raw materials, their classification and concepts defining properties relevant for guiding research and innovation has been elaborated.

3.2. The current raw materials playing field

3.2.1. Definition of the raw materials playing field. Afterwards, the current raw materials playing field and the players involved have been defined on the basis of two main tasks:

- the identification of private, public and multi-stakeholder platforms and formal networks with a focus on raw material supply, and

- the identification of current sectoral and horizontal policies at international, EU and member states level that might considerably affect the future availability of raw materials for the European industry at any point of the value chain (exploration, mining, production, trade, consumption, collection, reuse, remanufacturing, recycling). In particular, for sectoral policies the main sectors taken into account were: construction and infrastructures, energy (in relation to materials used for generation and storage facilities, CRM), water, ICT, chemistry, bio-economy (excluding energy and food).

The work of selecting, gathering, and highlighting information has been carried out by considering already existing review documents, such as those produced by the Department for Environment, Food and Rural Affairs of United Kingdom [11] or by the European Environment Agency [12], and coming also from EU projects, such as the MIN-GUIDE [13] or the Intraw [14] projects, to report just a few of them.

The results have highlighted that depending on the country, policies are addressing different aspects in relation to raw materials. However, at global level the common trend is focused on the willingness of accelerating the transition of economies towards more sustainability and circularity, especially for those policy bodies issued more recently. In relation to raw materials, this tendency is expressed by setting the necessary frameworks to become more competitive and leader in their respective domains (i.e. mining, refining, metallurgy, manufacturing, construction, recycling and other end of life management options). At national level, policies seek to achieve a better integration of sectoral policies (i.e. biomass across food, feedstock, energy, bio-chemicals sectors).

Exploration of the raw materials playing field. After the definition of the current situation, the exploration of the raw materials playing field has been conducted. The analysis has focused on both abiotic and biotic raw materials about: (i) ongoing funding activities, (ii) patents and papers of 27 large European enterprises belonging to 13 industrial sectors (Glass and Ceramics; Cement; Steel; Aluminium and Alloys; Catalysis; Fine Chemicals; Bio-Plastics; Pulp and Paper; Automotive; Aerospace; Tyres; Electronics and Consumers Products; Automation), (iii) EU business enterprise research and development expenditure (BERD), and (iv) (possibly planned) opportunities at international, EU and member states level for initiatives that:

- explicitly aim to enhance the availability of raw materials, and
- are likely to affect such availability, either by supporting research and development (R&D) activities or by facilitating upscaling and implementation.

Existing reports and/or studies on the effectivity and efficiency of EU and national funding programs have been analysed considering their impact on raw materials security of supply and the factors of success. Within the industrial sector, the consortium has mapped the current situation, trends, needs, long-term strategies, evolution and development aspects, oriented to grant the availability of raw materials at high level. Strategic aspects have been delivered in order to sum up in synergistic way to the research roadmaps and to highlight the congruence and differentiations. The raw materials R&I landscape has been analysed with regard to strength and weaknesses considering the coverage of sectors, raw materials and value chains.

The exploration of the raw material playing field has revealed that the current R&I programmes cover most relevant raw materials, industry sectors and value chain sections at EU and international levels. The R&D in the business enterprise sector in Europe is very heterogeneous and only three member states (Germany, France, and United Kingdom) hold 68% share of the total R&D. The research topics addressed in Europe and oversea are very similar, meaning that the international cooperation should be pushed in order to join efforts to solve common problems. Moreover, a more efficient R&D funding could be achieved if further funding for market access is provided.

3.2.2. The web-based portal. During all the activity of research, collection, and analysis of the current situation of EU raw materials sectors, a research and innovation raw material (R&I RM) portal has been created and implemented with all the collected data. The web-based portal, accessible at www.veram.eu, ensures the relevance and usefulness of the project results and constitutes a systematic, dynamic, open, and proactive platform for the entire stakeholder community, aiming to facilitate the sustainable coordination of the European R&I on raw materials. This platform defines and promotes the most adequate decision-making process/good governance and communication. Currently, the portal populates approximately 130 research projects, 1200 organizations, 75 funding agencies, 71 calls for proposals and 164 publications.

3.2.3. Possible future scenarios. The relevance of ongoing European R&D efforts has been investigated under three possible future scenarios (European stagnation, global scarcity and ethical value driven innovation) [15] for different material intensive industrial sectors and technology areas. These scenarios were identified through a four steps process, which includes:

- a brainstorming workshop, conducted within the Consortium, and the translation of the archetypical images of future [16] into the context of VERAM project by answering to the question "What do the architypes mean from the perspective of raw materials (e.g. demand, availability, development needs, industrial expectations)?";
- scenario formulation, applying the "Futures triangle method" [17,18], for which a plausible future (or futures) is created by analysing the interaction of the three forces: barriers (weight of history), drivers (push of the present), and compelling forces (pull of the future). Moreover, literature survey was carried out to strengthen the scenario storylines;
- scenario impacts were analysed in terms of opportunities and threats for both biotic and abiotic raw materials;
- finally, a report of the outcomes was written.

3.3. The 2050 Vision and Roadmap for European raw material research

3.3.1. The 2050 Vision for European raw material research. The results of both collection and analysis of data, together with the scenarios formulations, represented the first step for the development of the 2050 Vision and Roadmap for European raw material research.

From the background information, global outlook and regional/national ambitions coming from the gap analysis, the 2050 society and megatrends have been defined [19-21]. These global trends bring new societal challenges, that have the potential to boost the demand for products that make use of innovative technologies requiring R&I, such as the Key Enabling Technologies (KETs). In particular, KETs belonging to 65 innovation fields were considered for the VERAM project, since they are closely related to Raw Materials, belonging to 11 domains: Electronics and Communication Systems; Chemical Processes, Chemicals, Chemical Products and Materials; Manufacturing and Automation; Energy; Transport and Mobility; Construction; Mining, Quarrying and Extraction; Environment; Health and Healthcare; Textiles; Agri-food.

Matching all inputs, Europe by 2050 will be characterized by the transition towards a low-carbon economy, due to a number of circumstances, including climate change; population growth and the widening gap between social classes; the depletion of resources and the consequent demand of sustainable sourcing; and the mass customization and technological changes and innovations.

In order to face this future, the EU raw materials sectors will need to:

- provide and improve sustainable supply and use of primary, secondary and renewable RMs throughout the value chains;
- strengthen the EU economy by decreasing import dependencies and ensure base load supply trough diversification at source;
- embrace digitization, manage technical and practical applications of digitization in the sector, and revolutionize traditional value chains, business models and processes;
- develop and implement improved and better adapted measurable scientifically based values/indicators/standards associated with sustainability through the whole value chain where needs have been identified using newly developed data management systems;

- improve scientific and technical dialogue along and across business lines, exchanging experiences, advancing and leveraging good practices;
- increase the cross-disciplinary integration between academia and business for identification and development of new, cross-sectorial value chain opportunities.

For all the above-mentioned reasons, by 2050 it can be envisaged that the EU resources sector will:

- contribute to the gross domestic product and economic growth by providing EU market leadership in technology exports worldwide;
- contribute to the resilience of the EU industrial landscape and society through continually more advanced surveys of the EU landmass and marine environment, increasingly automated industrial processes, optimized valorisation of available resources, optimized use of resources and reduced consumption, increased recovery of materials and backflow into the economy and developing environmental footprint assessments of the full life-cycle of materials and products;
- addressing legacies and public acceptance through remediation of the EU's existing legacy sites for future use.

3.3.2. The 2050 Roadmap for European raw material research. On the basis of the results obtained by the gap analysis, the creation of the 2050 Vision, and the stakeholders' consultation, carried out through an online survey and focused on the collection of their personal vision for the raw materials sectors, a first draft of the 2050 Roadmap was produced.

The VERAM R&I Roadmap for 2050 comprises four pillars (already identified in the Vision document): supply, processing, closed loops and markets & applications. Each pillar is divided in different RIAs, for a total of ten. The content of the four pillars and the ten RIAs is reported below.

Pillar I (Supply) – Fostering a sustainable raw material supply to feed all value chains:

- RIA 1.1: New exploration and harvesting technologies for sustainable supply;
- RIA 1.2: Mobilizing an increased RM supply from EU sources.

Pillar II (Processing) – Resource efficient processing, refining and converting of raw materials:

- RIA 2.1: Development of resource efficient processing, refining and converting technologies;
- RIA 2.2: Minimization and valorisation of production residues.

Pillar III (Closed loops) – Maximizing material closed loops:

- RIA 3.1: Increasing collection, sorting, separation and detection efficiencies;
- RIA 3.2: Recycling technologies adapted to complex, durable, miniaturized and material efficient products;
- RIA 3.3: Developing and integrating assessment methodologies for balancing recycling costs and benefits.

Pillar IV (Markets & Applications) – Raw materials in new products and applications:

- RIA 4.1: Substitution of (critical) materials in new technology and energy applications;
- RIA 4.2: Development of new bio-based products;
- RIA 4.3: Development of new material applications and new markets.

The content of both pillars and RIAs are not definitive and will be updated and integrated by an interactive process that will endure throughout the project's life and that will ultimately produce the 2050 Roadmap for European raw materials.

4. Expected impacts

The outcomes of the VERAM project are expected to contribute towards the following impacts:

- develop a coordinated EU policy on raw materials research, innovation and technological developments to raise investor confidence and improve financing mechanisms;
- promote more efficient use of resources by preventing unnecessary overlaps in technology related to innovation activities;
- create cross-sectorial and cross value chain research through joint ETP, ERA-NETs and other stakeholders and national activities;

- provide opportunities for exchange of Best Practice amongst Member States;
- accelerate uptake of ground-breaking technologies;
- review research and innovation results achieved, forecast directions for future research and assist in reviewing the EIP Raw Materials Strategic Implementation Plan;
- identify synergies between individual initiatives at EU and national level;
- improve raw materials policy and access to their utilization;
- help reduce Europe's external supply dependency and foster a more efficient use of resources to improve the competitiveness of European industry;
- develop a common long term 2050 vision and R&I roadmap for European raw materials.

References

[1] Forest-based Sector Technology Platform 2013 [online] Brussels: Filip de Jaeger. Accessible at: http://www.forestplatform.org/#!/pages/6 [Accessed 14/10/2017]
[2] European Technology Platform for Sustainable Mineral Resources 2009 [online] Brussels. Accessible at: http://www.etpsmr.org/?post_documents=etp-smr-strategic-research-and-innovation-agenda [Accessed 14/10/2017]
[3] European Construction Technology Platform 2005 [online] Brussels. Accessible at: http://ectp.ectp.org/documentation/FA-MATERIALS-Vision2030-and-SRA-Version1-2005-09-02.pdf [Accessed 14/10/2017]
[4] The European Technology Platform for Advanced Engineering Materials and Technologies 2012 [online] Stuttagart: Steinbeis Advanced Risk Technologies. Accessible at: http://eumat.eu/filehandler.ashx?file=11580 [Accessed 14/10/2017]
[5] European Technology Platform for Sustainable Chemistry 2017 [online] Brussels. Accessible at: http://www.suschem.org/publications [Accessed 14/10/2017]
[6] Phaal R, Farrukh CJ and Probert DR 2004 *Technol. Forecasting Social Change* **71** 1
[7] Kostoff R and Schaller R 2001 *IEEE Trans. Eng. Manage.* **48** 2
[8] Ahlqvist T, Valovirta V and Loikkanen T 2012 *Sci. Public Policy* **39** 2
[9] Hussain M, Tapinos E and Knight L 2017 *Technol. Forecasting Social Change* **124**
[10] European Commission 2013 [online] Brussels. Accessible at: https://ec.europa.eu/growth/tools-databases/eip-raw-materials/ [Accessed 14/10/2017]
[11] Department for Environment, Food and Rural Affairs 2012 [online] London: Crown. Accessible at: https://www.gov.uk/government/publications/review-of-national-resource-strategies-and-research [Accessed 14/10/2017]
[12] European Environment Agency 2016 [online] Luxembourg: Publications Office of the European Union. Accessible at: http://www.eea.europa.eu/publications/more-from-less/ [Accessed 14/10/2017]
[13] http://www.min-guide.eu/ [Accessed 14/10/2017]
[14] http://intraw.eu/publications/ [Accessed 14/10/2017]
[15] Voros J 2003 *Foresight* **5** 3
[16] Dator J 2009 *J. Futures Stud* **14** 2
[17] Inayatullah S 2008 *Foresight* **10** 1
[18] Inayatullah S 2013 *There's a future: visions for a better world* (Madrid: BBVA Group) chapter 1 pp 37-65
[19] Geels FW 2002 *Res. Policy* **31** 8-9
[20] Geels FW and Schot J 2007 *Res. Policy* **36** 3
[21] Geels FW 2011 *Environ. Innovation Soc. Transitions* **1** 1

IOP Conf. Series: Materials Science and Engineering **329** (2018) 012002 doi:10.1088/1757-899X/329/1/012002

New life of recycled rare earth-oxides powders for lighting applications.

Pier Carlo Ricci [1*], Massimiliano Murgia [1], Carlo Maria Carbonaro [1], Serena Sgariotto[2], Luigi Stagi, [1] Riccardo Corpino [1], Daniele Chiriu [1], Maria Luisa Grilli [3]

[1] Dipartimento di Fisica – Università di Cagliari sp. N°8 Km 0.700, 09042 Monserrato, Cagliari, Italy

[2] Relight S.r.l., Via Lainate 98/100, Rho (Mi), Italy

[2] ENEA, Energy Technology Department, Casaccia Research Centre, Via Anguillarese 301, Rome, Italy

Abstract. In this work we analysed the optical and structural properties of Ce:YAG regenerated phosphors. The concentrate resulted as the final product of an industrial recycling process of waste electrical and electronic equipment (WEEE), and in particular fluorescent powders coming from spent lamps treatment plant. The waste pristine materials were re-utilized without any further purification and or separation process as starting materials to obtain a YAG matrix ($Y_2Al_5O_{12}$) doped with Cerium ions. We tested out the recovered concentrate against commercial Ce:YAG phosphors comparing their structural and optical properties by means of XRD measurements and steady time and time resolved luminescence. The analysis reveals that the new phosphors obtained by scrap powder have the same crystal structure as the commercial reference sample and comparable optical properties. In particular, the Ce-related emission efficiency has a quantum yield of about 0.75 when excited at 450 nm, in good agreement with our reference sample and with the one of commercial powder presently exploited in white LED. This achievement strongly suggests the possibility of a new life for the exhausted phosphors and a possible step forward to a complete circular process for lighting equipment.

* Corresponding author: carlo.ricci@dsf.unica.it

Introduction

Nowadays there is a spreading technological interest in the increase of efficiency of light sources. If one considers the revolution in lighting technology, both in concept, design and light systems, the society is dealing with in the late decades (from the first years of 2000's), the growing interest in the field is easily understood. According to McKinsey [1], the global market share of alternatives lighting sources as Light Emitting Devices (LEDs) and Compact Fluorescence Lights (CFLs), will raise from few percentage of the global market share in 2010 to almost 90% in the 2020. In particular, LED systems are foreseen to cover up to 70% of the lighting market, increasing from the current value of about US$26 billion (2017) to more than US$72 billion in 2020. The expected increase of LED exploitation has a great relevance in terms of environmental impacts, a reduction of

25% of traditional incandescent lamps by substitution with alternative sources corresponds to saving more than 180 million tons of CO_2 emitted.[1],[2]

The main drawback in LED and CFL technology is that LED and CFL systems make a large use of phosphors as emitting sources [3]. The latter are mainly represented by Rare Earths Elements (REE) based compounds, a requirement that can have potential impact on the economical point of view, due to potential supply disruption of raw materials and to elevated cost of their initial refining process. Currently, white emission in LED technologies (WLED) is achieved by using up or down conversion of phosphor emission. For WLEDs, phosphor down-conversion is most commonly based on a blue or near-ultraviolet LED combined with a yellow-emitting phosphor, or a combination of different phosphors to produce a broad energy distribution [4]. The phosphor(s) may be incorporated into the LED package, or may be located remotely. Fluorescent lamps (CFL) also exploit phosphors' down-conversion, the fluorescent powders being deposited on the lamp bulb and excited by the emission of the mercury vapour. In both the cases, Ce doped Yttrium Aluminium Garnet (Ce:YAG) represents the phosphor of choice for its spectral properties. The ubiquity of Ce:YAG phosphors in commercial devices is related to their very simple and cost-effective fabrication, as well as their long term stability, which is related to the high melting point and stable thermal and chemical properties of the system. However, the supply of REE, such as Ce and Y in Ce:YAG phosphors, to European industry completely relies on China, the main world producer of REE compounds, and such a dependence of the supply chain represents a potential disruptive economic threat. Indeed, the concentrated supply of REE, where a portion of about 99 % and 87 % for heavy and light REE, respectively, are currently sourced from China, has resulted in a supply chain for European end-users that is not sufficiently robust when considering high capital investments in modern technologies. [3] Although different alternatives to RE based phosphors were already proposed to Ce:YAG phosphors (organics, hybrids, nanodots) [5]-[14], the use of rare earths seems still with the higher potential solution to easy implementation in solid state lighting devices.
In order to reduce this threatening dependence on China REE supply, European industry must develop novel "REE-free" technologies and/or reduce the REE waste and increase its capability to re-use and recycle them. An archetypal example in this respect is the supply chain of Aluminium in Italy, where no mines of raw materials are present and the whole national production of Al relies on recycling politics. The aim of this paper is to demonstrate a possible low cost way to exploit REE waste of CFL as secondary source of phosphors.

LED waste is classified by the 2012/19/EU Directive as a Waste Electrical and Electronic Equipment (WEEE). The recycling processes, widely applied by the European recyclers, are mainly addressed at the recovery of major commodity fractions (e.g. plastics, glass), with still high amounts of landfilled materials. However, the presently un-recycled residual fraction contains many strategic elements at concentration even higher than primary ores like Rare Earths (such as Yttrium, Lutetium, and Europium). Consequently, the current management strategy of end of life LED/CFL lighting devices determines two potentially extremely negative impacts, a dangerous contamination of the environment and the neglected extraction of critical raw materials. The methods for Yttrium and other REE extraction, described by patents and scientific articles [15]-[17], are based on hydrometallurgical and pyrometallurgical processes. Pyrometallurgical extraction requires high consumption of energy, has very high environmental impact due to gaseous emissions and requests high capital investments, so that being unsustainable for the European SMEs recyclers. For this reason hydrometallurgical processes are generally preferred, but there are still a few drawbacks, the main one for Yttrium, REE and Indium recovery from target waste being the too high energy consumption for the production of high purity products.

In this scenario, the present work deals with the possibility to re-use the Ce:YAG fine fraction residue of the hydrometallurgical treatment from spent lamps without any further chemical purification/separation or extraction as starting material to obtain new phosphors for WLED emission, thus promoting a low cost process for secondary sources' supply with virtual no impact on the environment.

Experimental

Two sets of Cerium doped YAG samples have been prepared; for the preparation of set 1 (YAG1), waste powders from the industrial plant of the Relight srl have been utilized.[18] The chemical composition is reported in table 1, evidencing the large contribution of Yttrium oxide (86.12%), and minor contributions of Europium oxide (5.98%) and Gadolinium oxide (2.98%).
The starting oxides were dissolved in nitric acid to convert them into corresponding nitrates. The powder was dissolved in 600 ml of 0,4 mol/l solution of the Nitric acid. The solution was agitated for 8 hours at temperature of 50 °C. The pH was monitored and, when needed to preserve pH in the range of 4.5-5.0, concentrated nitric acid was added.
Since in the starting oxides there was an excess of europium oxide and a lack of aluminium and cerium ions, pure $Al(NO_3)_3 \cdot 9H_2O$ (Aldrich), and $Ce(NO_3)3 \cdot 6H_2O$ (Aldrich) was added in the stoichiometric amount to obtain the right molar ratio.
The second set (YAG2) was obtained starting from pure nitrates; Pure $Al(NO_3)_3 \cdot 9H_2O$ (Aldrich), $Y(NO_3)_3 \cdot 6H_2O$ (Aldrich) and Ce(NO3)3·6H2O (Aldrich) were utilized.
For both the synthesis the nitrate salts were dissolved in 80 ml of deionized water and acetic acid was then added as complexing agent. These metal salts were dissolved at a molar ratio, Y:Al, of 3:5, respectively. Ce was added to obtain a concentration of 1.0 at.%.
Each solution was heated with stirring at 60°C for 1 hour. Then, ethylene glycol was added to the solution until it became a gel. Successively it was dried at about 100°C for three days and calcined at 1250 °C, each sample for 12 hours.
X-ray diffraction patterns were collected by using a Seifert X3000 diffractometer with a θ–θ Bragg-Brentano geometry with Cu-Ka wavelength.
Steady state photoluminescence (SSPL) measurements were performed using the filtered light from a laser driven Xenon lamp (EQ- 99-X) with a final bandwidth of about 1 nm, while time resolved luminescence (TRPL) measurements were performed using an optical parametric oscillator with a frequency doubler device, pumped by the third harmonic (355 nm) of a pulsed Nd:YAG laser (Quanta-Ray Pro 730). The excitation pulse width at half-maximum was 8 ns with 10Hz of repetition rate, and spectral bandwidth less than 0.3 cm 1. The collected signal was dispersed by a spectrograph (ARC-SpectraPro 300i) with a spectral bandpass <2.5 nm and detected by a gateable intensified CCD (PI MAX Princeton Inst.). The preparation of a white LED from waste powders has been obtained by drop casting a suspension of the YAG1 powder in a methanol solution over a blue LED (OSRAM Golden DRAGON, LD W5SM) with main emission at 460 nm.

Table 1 atomic percentage of REE oxides in scrap powders. The scheme of the extraction process of the REE concentrate is also reported on the left

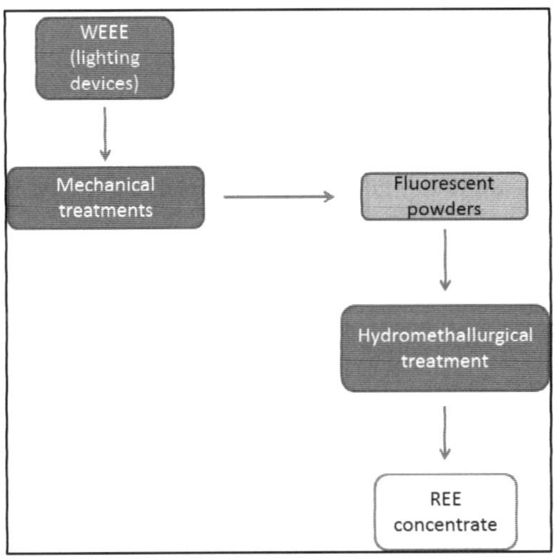

Material	atomic content (%)
Lanthanum oxide	0.02
Cerium oxide	0.96
Europium oxide	5.98
Terbium oxide	0.73
Yttrium oxide	86.12
Gadolinium oxide	2.98
others	3.21

Results and Discussion

Figure 1 reports the XRD diffraction pattern of the Ce:YAG samples obtained from scraps (YAG1) and from pure/commercial starting materials (YAG2). The spectra are in agreement with the JCPDS card: 79-1892/1 relative to the YAG structure. Low intensities peaks at about 2θ= 28° e 47° reveals the presence of minor phase impurities, most probably $Al_2Y_4O_9$ in both the samples, due to the relative low temperature synthesis. It is worth nothing that minor peaks related to the presence Y_2O_3, $YAlO_3$ and Ce_2O_3 are observed.
The Scherrer analysis has been performed to have an estimation of the average dimension of the nanocrystals. According to the equation:

$$D = \frac{0.89K_\alpha}{\beta\cos\theta}$$

where β is the line broadening at half maximum in radians observed in the experimental spectrum at 2θ = 18.10° and K_α=0.1540 nm. The average dimension of the nanocrystals can be estimated for the two set of samples between 40 and 50 nm.

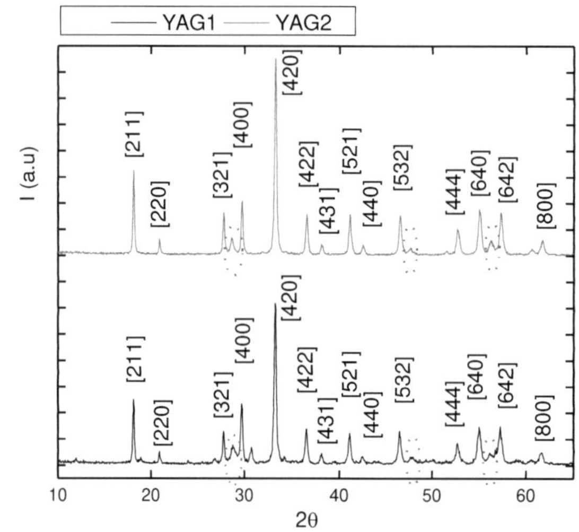

Figure 1 - XRD diffraction pattern of sample YAG1 (obtained from scraps) and YAG2 (from pure starting compounds). Peak indexes of the YAG phase are indicated, the peaks from secondary phases are indicated with green dotted circles

The emission spectra of the two samples, excited at 450 nm, are almost overlapped, a further indication of the close resemblance of the crystallographic structure (**Figure 2**) of the two set of samples because of the well-known effect on the emission properties of the crystal field (see in following). Indeed, we observed only a small broadening of the emission band towards the lower spectral energy side in scrap YAG1 sample as compared to our YAG2 reference. The emission band peaked at about 450 nm nm arises from Ce^{3+} ions and is related to the recombinations from the 5d levels to the fundamental doubly split 4f levels $^2F_{5/2}$ and $^2F_{7/2}$ [19].Generally, the effect of local crystal field surrounding the Ce^{3+} ion affects the position of the 5d levels, while is rather weak on the 4f states due to the shielding effect of the outer 5p and 6s electrons [19]. Since the starting scrap powders of YAG1 contain Gd oxide and Ce doped Gadolinium Aluminium Garnet could be formed during the synthesis (even though at a concentration level below the sensitivity threshold of XRD measurements). Indeed, the emission of Ce^{3+} ions is redshifted in Ce:GdAG as compared to Ce:YAG [20],[21].

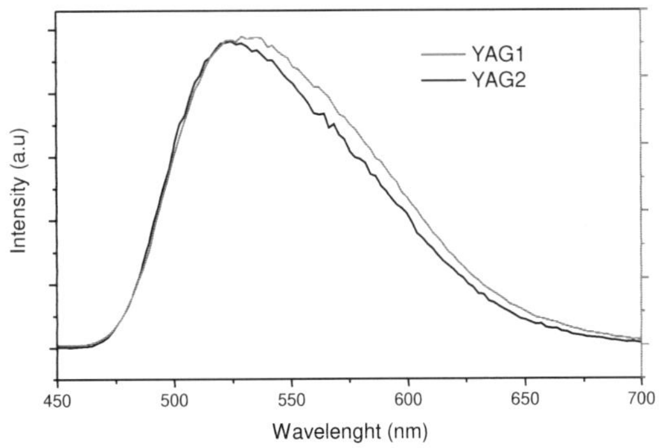

Figure 2 - Steady time photoluminescence spectra of YAG1 and YAG2 samples. Excitation wavelength: 450 nm

Despite the presence of Ce:GdAG phase and the low quantum efficiency of radiative recombinations of Ce in Ce:GdAG, no substantial decrease of efficiency in recovered YAG1 samples was observed with respect to our reference sample (YAG2). The quantum yield of the emission was evaluated according to the following equation [22]:

$$QY = QY_R \frac{I}{I_R} \frac{OD_R}{OD} \frac{n^2}{n_R^2}$$

where the subscript R refers to our reference (Coumarin 540, C6), I is the PL intensity excited at 450 nm, OD indicates the optical density (that is the absorbance) which was evaluated at 450 nm, and n refers to the refractive index (at 450 nm) of the samples (ethanol for the C6 reference solution) [22].
We estimated a quantum efficiency of about 0.75 for both the samples, gathering a 17% decrease as compared to commercial bulk samples, where the QY is of about 0.9 [19], [23]. As we discuss in the following, we ascribe the reported decrease of QY to the nanosized dimensions of our samples, instead of to the presence of impurities or defects.

To go further in the characterization of the samples, we recorded time resolved luminescence of both the samples excited at 450 nm. The spectra, reported in **Figure 3**, and the time decays monitored at the peak position (550 nm) are very similar for both the samples. We also verified that the spectral profile did not change when recorded at different delays from the excitation pulse down to 150 ns time delay. This is a further proof that the recombinations take place from the same Ce^{3+} emitting site and no secondary emissions and/or charging phenomena to other centers are observed.

The decay kinetics can be well represented by a multiexponential decay model by assuming two different decays. In this model, the PL intensity at a certain time *t* is the sum of individual single-exponential decays[22]:

$$I(t) = \sum_{i=1}^{n} \alpha_i \times e^{-t/\tau_i}$$

where τ_i are the decay times and α_i their amplitudes at time t = 0 (n is the number of decays, n=2 in the present case).

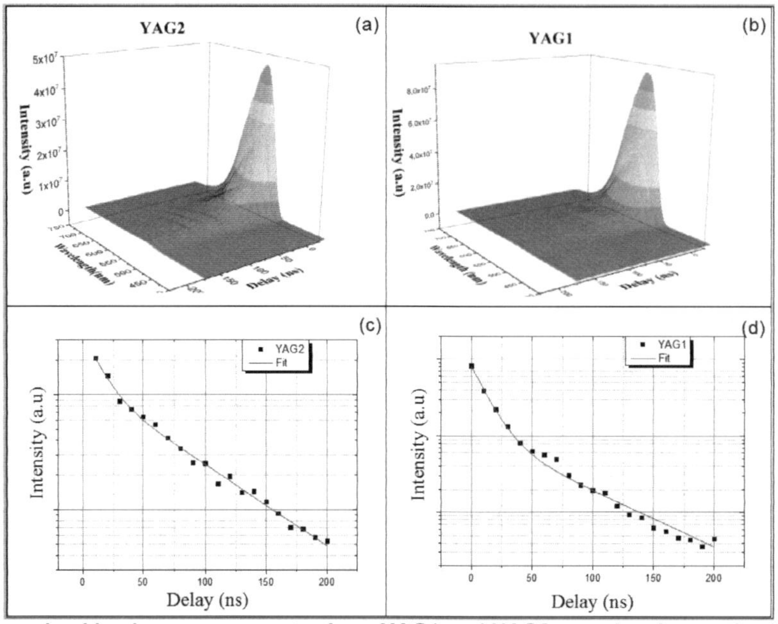

Figure 3 Time resolved luminescence spectra from YAG1 and YAG2 samples. In panels a) and b) the full decay spectra are reported; In panels b) and c) the time decay evolution is reported monitored at 550 nm. Excitation wavelength: 450 nm

The presence of two time decays is indicative of two recombination paths, with estimated time decays of 8 ns and 65 ns for the faster and slower component respectively (**Figure 3**). Since the spectral profile did not change during the decay, we can hypothesize that there are two different mechanisms of depletion of the emitting centers. The slow decay is in good agreement with literature data of Ce^{3+} decays of about 60-70 ns in bulk YAG matrices. Typically, this time decay is well represented by a single exponential decay [23]. In our case, however, we have two decays, and the second fast one could be related to some non-radiative channels depleting the Ce^{3+} excited level. Therefore we propose to assign the slow recombination channel of our samples to "bulk" recombination and the faster one to recombination at Ce^{3+} sites close to the surface. The proposed assignment is supported by the estimated nanosize dimensions of our samples, being the powders constituted by nanoparticles with mean average diameter of about 45 nm. At nanoscale, due to the surface-to-volume ratio, the contribution of surface centers, including both radiative and non-radiative paths, became relevant. The non-radiative path can be related to the presence of OH group at the surface, or, more generally, to the presence of surface defects. Another possibility is the formation of Cerium clusters at the surface, related to the higher ionic radius of Ce^{3+} (1.034 Å) with respect to Y^{3+} (0.880 Å). We would like to point out that the mechanism leading to the presence of a faster decay, responsible of the smaller QY estimated with respect to bulk samples, is the preparation procedure that settles the nanosized dimensions of our crystals. Indeed, we retrieved the same time decay plot in both the samples, the scrap recovered YAG1 and our reference YAG2, both prepared with the same procedure described in the experimental section.

To evaluate the weight of the surface contribution we estimated the fractional contribution f_i of each time decay. It also represents the contribution of each component to the steady-state intensity, allowing comparing time resolved and steady state spectra. The fractional contribution is defined as follows [22]:

$$f_i = \frac{\alpha_i \tau_i}{\sum_j \alpha_j \tau_j}$$

As extracted from the previous equation, the fractional contribution of the fast surface de-excitation accounts for 28% and 20% of the overall emission of YAG1 and YAG2, respectively. We also estimated the average lifetime <*t*>, defined as a weighted mean of gathered data by assuming a multiexponential decay model [22], where the contribution of each decay to the average decay is weighted. The average lifetime <*t*>, for a double-exponential decay, can be defined accordingly:

$$\langle t \rangle = \frac{\alpha_1 \tau_1^2 + \alpha_2 \tau_2^2}{\alpha_1 \tau_1 + \alpha_2 \tau_2} = f_1 \tau_1 + f_2 \tau_2$$

The data indicate that there is a very slightly difference among YAG1 and YAG2 with average lifetime of 48 ns and 54 ns, respectively.

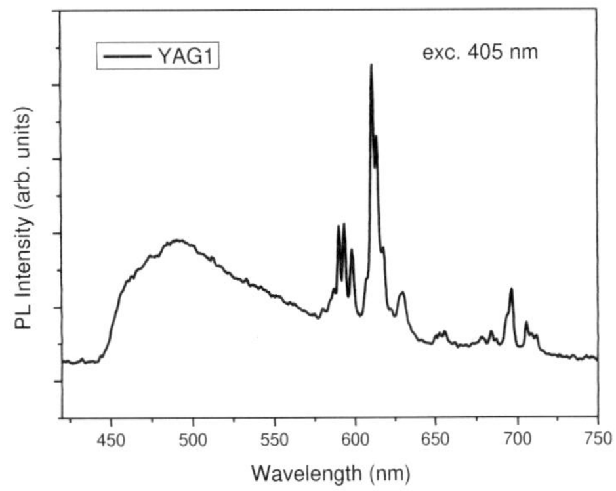

Figure 4 Steady time photoluminescence spectrum of YAG1 sample. Excitation wavelength: 405 nm.

To characterize further the scrap powders, we investigated the possible presence of impurities by exciting at 405 nm. Indeed, aiming to exploit our recycled powders as possible secondary source of phosphors in lighting applications, the typical excitation is in the 400-450 nm range. When exciting at smaller wavelengths, the emission contribution of possible impurities can be detected. It is worth nothing that the presence of impurities is not always detrimental, but can be exploited to change the emission properties of the phosphors. Actually, by exciting YAG1 sample at higher energy, besides the recombination from Ce^{3+} ions, still observed but with decreased efficiency, we gathered the emission from Eu^{3+} ions. Those ions are embedded in the YAG1 matrix but could not be excited a larger wavelengths. In this framework, **Figure 4** reports the emission spectra excited at 405 nm, where additional narrow bands with respect to the broad Cerium emission are clearly evident. The main peaks, centered at 590 nm, 610 nm and 710 nm, are assigned to the recombination $^5D \rightarrow {}^7F_j$, j=1,2 and 4 respectively of Eu^{3+}.

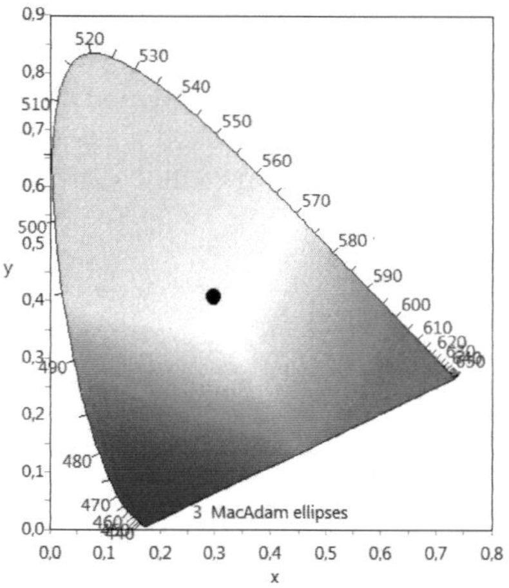

Figure 5 - CIE diagrams of YAG1 sample. The spectrum contains also the residual component of the 460 nm line of a commercial LED

Finally, in order to prove the effective possibility to reutilize the obtained phosphor, a blue led (emission at 460 nm) was covered with a thin layer of YAG1 by drop casting the powder dispersed in ethanol solution. As expected, the emitted spectra presents the emission of the YAG1 phosphors coupled with a strong residual component of the electro-emitted radiation by the LED (CIE coordinates: x=0.3041, y=0.4095). The CIE spectrum, reported in **Figure 5**, evidences the perceived white light, and indicates the way we can re-use the waste of CFL or display systems.

Conclusions

In this work we demonstrated that Ce:YAG regenerated phosphors synthesized starting from scrap powders derived by an industrial recycling process of waste electrical and electronic equipment from exhausted CFL can find new applications in lighting field as secondary sources of phosphors. The samples were synthesized by following a nitrate-based low temperature procedures that allows to prepare crystalline nanoparticles of about 45 nm in diameter. The regenerated powders have a quite high quantum yield (0.75), only 17% lower than commercial bulk samples, and comparable to the one of samples produced by the same synthesis but starting from commercial powders. We realized a white LED device by drop casting the powders dispersed in ethanol on a commercial blue LED. The optimal optical performances and the good quality of the white light perceived in a model lighting application strongly suggest the possibility of a new life for the waste regenerated phosphors and a possible step forward to a complete circular process for displays and CFL

References

[1] McKinsey & Company Lighting the Way: *Perspectives on the Global Lighting Market* 2nd edn; (2012);
[2] Evstratyeva, K Market *Overview for Global LED Industry: 2013–2018* (Strategies Unlimited/Pennwell, 2014).
[3] Report *"Critical raw materials for the EU" by ad-hoc working group on defining critical raw materials of Raw Materials Supply Group*. May 2014- http://ec.europa.eu/enterprise/policies/raw-materials/critical/index_en.htm
[4] National Research Council. 2013. *Assessment of Advanced Solid-State Lighting*. Washington, DC: The National Academies Press.
[5] Luridiana, A, Pretta, G, Chiriu, D, Carbonaro, CM, Corpino, R, Secci, F, Frongia, A, Stagi, L, Ricci, PC 2016 *RSC Advances*, 6 (26), pp. 22111-22120
[6] Chiriu, D, Stagi, L, Carbonaro, CM, Ricci, PC 2016 *Physica Status Solidi (C) Current Topics in Solid State Physics,* 13 (10-12)
[7] Chiriu, D, Stagi, L, Carbonaro, CM, Ricci, PC 2016 *Physica Status Solidi (C) Current Topics in Solid State Physics,* 13 (10-12), pp. 989-997.
[8] Chiriu, D, Stagi, L, Carbonaro, CM, Corpino, R, Casula, M.F., Ricci, PC 2016 *Materials Research Bulletin,* 77, pp. 15-22
[9] Carbonaro CM,, Ricci PC, Grandi S, Marceddu M, Corpino R, Salis M, Anedda A, 2012 *RSC Advances*, 2 (5), pp. 1905-1912
[10] Carbonaro, C.M., Orrù, F., Ricci, P.C., Ardu, A., Corpino, R., Chiriu, D., Angius, F., Mura, A., Cannas, C. 2016 *Microporous and Mesoporous Materials* 225,pp 432-439
[11] Ricci, PC, Carbonaro CM, Casu , A, Cannas C, Corpino R , Stagi L,. Anedda A, 2011 *J. Mater. Chem.* 21 (21), pp. 7771-7776 .
[12] Kim, T-H, Chung, D-Y, Ku, J, Song, I, Sul, S, Kim, D-H, Cho, K.-S., Choi, B.L., Min Kim, J., Hwang, S., Kim, K. 2013 *Nature Commun.* 24, 2637 .
[13] Choi, MK, Yang, J, Kang, K, Kim, DC, Choi, C., Park, C, Kim, SJ, Chae, SI, Kim, T-H, Kim, JH, Hyeon, T, Kim, D-H 2015 *Nature Comm.* , 6, art. no. 7149.
[14] Wei, W, Xu, H, You, Q, Cheng, Q, Liu, C, Zou, L, Liu, X, Liu, J, Cao, Y-C, Zheng, G 2016 *Mater and Design*, 91, pp. 165-170
[15] Beolchini F, Fonti V, Dell'anno A, Rocchetti L, Vegliò F 2012 *Waste manage* 32, 949-956
[16] Cui J, Zhang L. 2008 *J. Hazard Mater* 158, 228-256
[17] Zhifeng L Patent (2015) CN103191903 (B) "Method for recycling waste LCD (liquid crystal display) panels"
[18] Innocenzi, V Michelis, ID, Sgarioto, S,Gotta, D, Kopacek, B,Veglio, F 2016, Electronics goes green 2016+, 7829831.
[19] He, X Liu X, Li R, Yang B, Yu K, Zeng M, Yu R 2016 *Sci. Rep.* ; 6: 22238.
[20] Joshi C, Yadav P, Moharil SV 2014 *Trans. Electrical and electronics Materials*, 15, 69-72
[21] Tien T Y, Gibbons EF, DeLosh RG, Zacmanidis P J, Smith DE, and H. Stadler L 1973 *J. Electrochem. Soc.,* 120, 278
[22] Lakowicz JR, Principles of Fluorescence Spectroscopy, 2nd Ed., Kluwer Academic, 1999
[23] Gorrotxategi P, Consonni M, Gasse A 2015 *J. Solid State Light.* 2:1

Life cycle assessment of metal alloys for structural applications

K Malovrh Rebec[1], B Markoli[2] and B Leskovar[2]

[1] Slovenian national building and civil engineering institute, Department of Building Physics, Ljubljana, Slovenia
[2] University of Ljubljana, Faculty of Natural Sciences and Engineering, Department of Materials and Metallurgy, Ljubljana, Slovenia

Abstract. The study compared environmental footprints of two types of Al-alloys: well-known 5083 aluminium alloy with magnesium and traces of manganese and chromium in its composition. This material is highly resistant to seawater corrosion and the influence of industrial chemicals. Furthermore, it retains exceptional strength after welding. The comparisons were made to an innovative alloy where the aluminium based matrix is reinforced by metastable quasicrystals (QC), thus avoiding magnesium in its composition. Furthermore, we checked other aluminium ingots' footprints and compared European average and Germany country specific production data. Environmental footprints were assessed via cradle to gate life cycle assessment. Our findings normalized to 1 m^2 plate suggest, that newly proposed alloy could save around 50 % in value of parameters abiotic resources depletion of fossil fuels, acidification, eutrophication, global warming potential and photochemical ozone creation potential if we compare Qc5 to 6 mm 5083 alloy plate. Only abiotic resources depletion of elements and ozone depletion parameters increase for Qc5 compared to 6 mm 5083 alloy plate.

1. Introduction

1.1. The aim of the study
This study has focused on environmental impacts of aluminium alloys known for their high specific strengths and rigidity. Indispensable in a wide variety of applications such as skins of jets and spacecraft fuel tanks, storage tanks, railcars, vehicles, ships and others, Al-alloys are part of aerospace, military and transportation product manufacturing. The study compared two types of Al-alloys: well-known 5083 aluminium alloy with magnesium and traces of manganese and chromium in its composition. This material is highly resistant to seawater corrosion and the influence of industrial chemicals. Furthermore, it retains exceptional strength after welding [1]. The comparisons were made to an innovative alloy where the aluminium based matrix is reinforced by metastable quasicrystalline phases (QC). Quasicrystals have many interesting properties such as high hardness, thermal stability, high corrosion resistance, low coefficient of friction and low interfacial energy [2, 3]. High symmetry and quasiperiodicity contribute to forming the orientation relationship on all the interfaces in the bulk. This leads to important properties such as significant enhancement of strength and ductility of alloys containing quasicrystals [4-6].

1.2. Assumptions made in the study
The study took into account different thicknesses of plates since their characteristics vary in terms of specific strength and rigidity (see Table 1). For comparison, typical 5xxx series alloy 5083-O in soft

(annealed) state and Al-alloy reinforced with QC particles Qc5 in the as-cast state were analysed. The 5083-O alloy has the ultimate tensile strength of 290 MPa, tensile yield strength of 140 MPa and elongation at break of 22 % [1], while Qc5 alloy reinforced with QC particles has ultimate compression strength of 610 MPa, compression yield strength of 280 MPa and contraction at break of 32 %. If we assume that the difference between compression and tensile measurement of yield strength is negligible, one can calculate (linear dependence) that alloy reinforced with QC particles will have the same yield strength as 5083-O, when the diameter of the novel alloy is reduced by 50 %. Our observed unit was 1 m^2 panel with similar mechanical characteristics when integrated in applications. However, comparisons were done on 1 kg basis as well, since previous studies presented data in this manner.

Table 1. Plates characteristics.

	5083	Qc5
Plate thickness	6 mm	3.9 mm
Density	2.73	3.20
Volume	6000	3900
Plate mass	16.38	12.20

1.3. System boundaries

The modular principle described in the standard EN 15804 has been applied for the system boundaries of this life cycle assessment [8]. The assessment relates to A1 to A3 modules, this is the production phase (from cradle to gate). We have included raw material extraction and processing background data, processing energy data and secondary material input (see **Figure 1**). Transport was not included in the study.

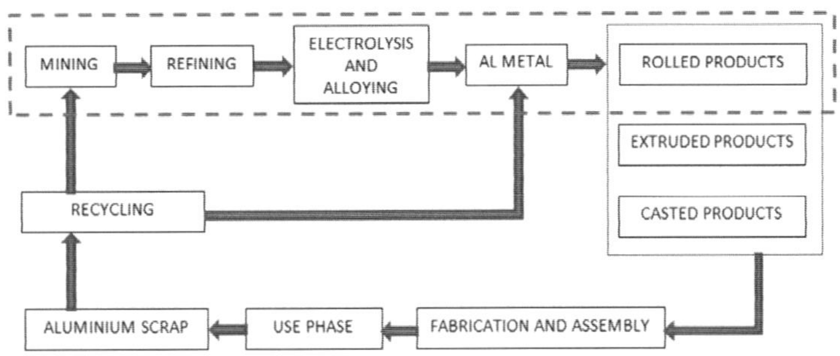

Figure 1. System boundaries.

Life cycle assessment (LCA) followed the rules in ISO 14040 and 14044 (procedure) [7, 9]. Thus principles definition of the goal and scope of the LCA, the life cycle inventory analysis (LCI) phase, the life cycle impact assessment (LCIA) phase, the life cycle interpretation phase, reporting and critical review of the LCA, limitations of the LCA were applied. Pilot production on a laboratory scale provided limited data, thus some estimates were taken into account.

2. Assessment details

2.1. Production process and materials bill

Novel Al-alloys reinforced with QCs can be classified as aluminium matrix composites (AMCs). AMCs consist of at least two chemically and physically distinct phases that are suitably distributed to provide properties which are not obtained with either of the individual phases. The reinforcing material is usually in the form of the fibres, particles of flakes, while the Al matrix phase is generally continuous and ensures the compliant support. Reinforcing phase, stronger than the matrix, is embedded in the matrix in discontinues way [10]. In our case, a matrix is aluminium based and the reinforcement phase is a metastable QC phase. The production process is thus majorly simplified by the fact, that reinforcing QC phase grow from within the matrix during the solidification of the melt. The investigated aluminium alloy sheets were cast to a thickness ranging between 3.9 and 7 mm by twin-roll strip casting technology. Cold rolling the strips to an intermediate gauge and reroll annealing the intermediate gauge material is then applied. The reroll-annealed material is then cold rolled to a final sheet gauge followed by a final recrystallizing or back annealing. The combination of controlled casting parameters, controlled amounts of iron, silicon, manganese (Mn), chromium and magnesium (Mg) and reroll and final annealing temperatures results in an improved sheet product in terms of finer grain size, higher elongation and formability, age softening and better corrosion resistance.

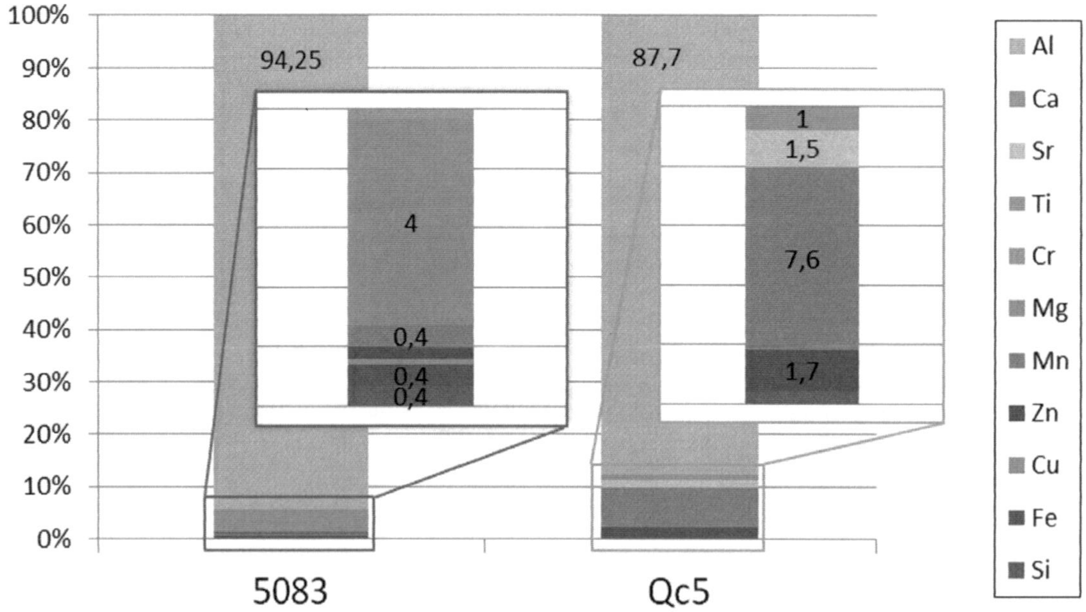

Figure 2. Materials composition (%) for 5083 alloy (left) and Qc5 alloy (right).

2.2. Impact assessment

We have analysed environmental impacts for the production phase of rolled products (see Figure 2). The boundaries were thus set a bit differently than in some other studies, where recycling has also been assessed, for example in a life cycle assessment report published by Aluminum Association [11]. The focus of our research was mainly to compare different compositions of alloys. The results were calculated according to the standard EN 15804, chapter 7.2.2 [8]. The impact assessment was carried out for the impact categories using CML 2001 characterisation factors. We will report and compare the following parameters in this presentation: Global warming potential [kg CO_2-Equiv.], Ozone depletion potential [kg R11-Equiv.]; Acidification of land and water potential [kg SO_2-Equiv.]; Eutrophication potential [kg Phosphate-Equiv.]; Photochemical ozone creation potential [kg Ethene-Equiv.]; Depletion of abiotic resources potential (elements) [kg Sb-Equiv.] and Depletion of abiotic resources potential (fossil) [MJ]. The calculations were done in Gabi software (Thinkstep).

Global warming potential (GWP_{100}) is a relative measure of how much heat a greenhouse gas traps in the atmosphere. It compares the amount of heat trapped by a certain mass of the gas in question to the amount of heat trapped by a similar mass of carbon dioxide. A GWP is calculated over a specific time interval (100 years). GWP is expressed as a factor of carbon dioxide (whose GWP is standardized to 1). **Acidification Potential (AP)** caused by pollutants SO_2, NO_x, NH_x affect the soil, groundwater, surface water, biological organisms, ecosystems and materials (buildings), which is one of the consequences of crumbling building materials and accelerated corrosion of surfaces. **Eutrophication Potential (EP)** measures the increase of excess nutrients in the water (especially nitrogen and phosphorus). Excess nutrients alter the balance of species in aquatic and terrestrial ecosystems. Large concentrations of excess nutrients can also seriously affect drinkability of surface waters. **Ozone Depletion Potential (ODP)** as a result of anthropogenic emissions. Thinning of the ozone in the stratosphere increases the proportion of UVB radiation that reaches the Earth's surface. UVB radiation relates to human and animal health, the balance of aquatic and terrestrial ecosystems, biochemical cycles and materials. **Photochemical ozone creation potential (POCP)** - the formation of ozone in the troposphere (bordering the Earth's surface) is caused by oxides of nitrogen (NO_x) and volatile organic compounds (VOC). Ozone in the troposphere causes damage to the materials (accelerated oxidation), plants (crop injury) and has adverse effects on humans (eg. Asthma). **Abiotic Resources Depletion Potential** - fossil fuels and other sources are evaluated separately.

2.3. Estimations and background check

We have found previous studies and data of other aluminium alloys A1-A3 environmental footprints and compared them. All comparisons in this chapter were normalized to 1 kg of material. First, we have checked country specifics and found out the aluminium ingot mix footprint for Germany, for example, causes roughly 9 kg CO_2-Equiv in terms of global warming potential, but EU average is a bit lower, 8.3 CO_2-Equiv. This is probably due to country-specific production energy background environmental data, since the material composition for those two cases is the same.

Then, we have found and compared environmental data for aluminium sheets of different material composition: $AlCu_4Mg$, $AlMg_3Mn$ and $AlMg_3$ all with the Germany country-specific data. Furthermore, a very interesting study by Nuss and Eckelman presented a Periodic table of global warming potentials (GWPs) and the data we took into account are consistent with the results of the mentioned study [12].

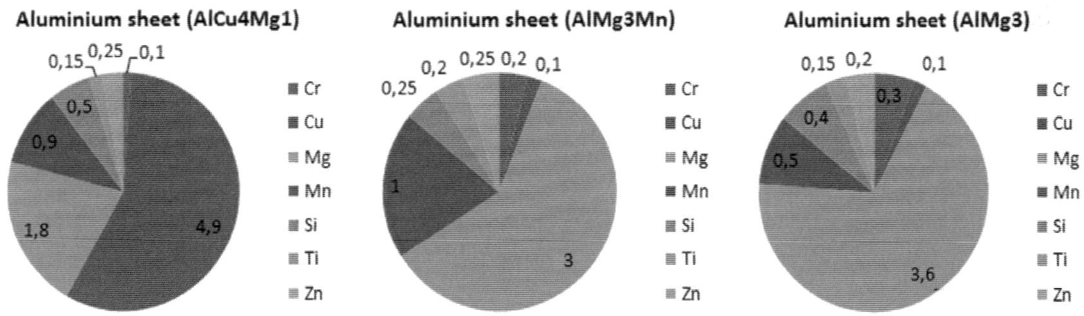

Figure 3. Materials composition of alloys.

Our conclusion was that material composition is an important factor as is energy and auxiliary materials used in production.

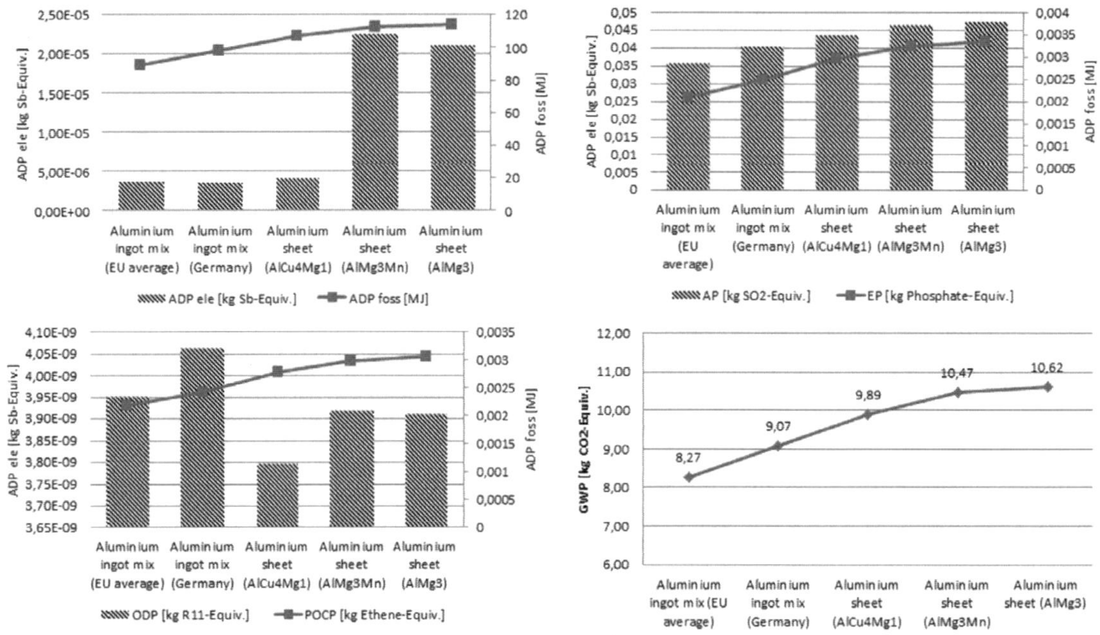

Figure 4. Results of previous studies: different alloys and comparison between European average and German country-specific data.

3. Results

3.1. Innovative Qc5 material calculation rules, datasets and system background

Cut-off rules took into account the fact, that only pilot laboratory production data is available at the moment for innovative metal alloy. LCA analysis included all input data regarding materials and their specifics. Basic raw material and auxiliary materials (water) were modelled by datasets provided in by Thinkstep and Ecoinvent. Consisting of various processes in which the production or recovery of multiple metals occurs simultaneously, the metals production system is highly interconnected. For the innovative alloy production, dataset for Slovenian electricity mix has been applied in the model. Process energy has been summed up and imitated by dataset "Thermal energy from natural gas" (thus a safety factor has been taken into account). Compared alloys differ in materials composition, 5083 alloy consists of 94 % of aluminium, on the other hand, aluminium represents only around 88 % of the total mass in the innovative Qc5 alloy. 7.6 % manganese is added to Qc5 alloy and no magnesium. But in 5083 there is 4 % of magnesium and only 0.4 % of manganese. Thus, we have expected the significantly lower environmental impact in terms of materials composition.

3.2. Results and interpretation

Results were calculated per 1 kg of alloy and for 1 m^2 of the plate so we were able to compare them to alloys investigated in the previous studies.

Table 2. Results for compared alloys (5083 and innovative Qc5).

		per 1 kg	per 1 kg	per 1 m^2	per 1 m^2
ALLOY		**5083**	**Qc5**	**5083**	**Qc5**
ADP ele	[kg Sb-Equiv.]	1,07E-05	2,46E-05	1,74E-04	3,00E-04
ADP foss	[MJ]	9,89E+01	9,08E+01	1,62E+03	1,11E+03
AP	[kg SO2-Equiv.]	4,23E-02	4,38E-02	6,93E-01	5,35E-01
EP	[kg Phosphate-Equiv.]	2,91E-03	2,50E-03	4,77E-02	3,05E-02
GWP	[kg CO2-Equiv.]	9,37E+00	8,45E+00	1,53E+02	1,03E+02
ODP	[kg R11-Equiv.]	4,58E-09	8,96E-09	7,50E-08	1,09E-07
POCP	[kg Ethene-Equiv.]	2,74E-03	2,54E-03	4,49E-02	3,10E-02

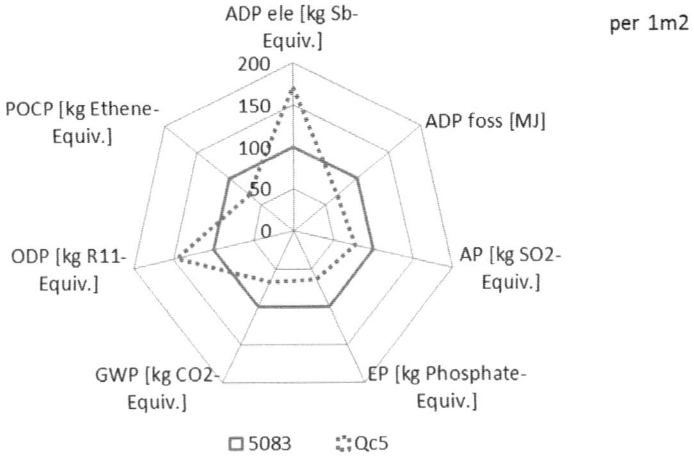

Figure 5. Comparison of the results for 1 m^2 plates of 5083 alloy and innovative Qc5 alloy in % of multiple environmental footprint potential parameters.

Our general finding include are that the innovative alloy Qc5 has significantly smaller environmental footprints in terms of global warming potential, eutrophication potential, acidification potential, fossil abiotic depletion potential and photochemical ozone creation potential per 1 m^2 plate compared to 5083 alloy. However, ozone depletion potential and elements abiotic depletion potential are higher than the footprints of 5083 alloy 1 m^2 plate. Results revealed 33 % lower global warming potential, 33 % lower eutrophication potential, 23 % lower acidification potential, 32 % lower fossil abiotic depletion potential and 31 % lower photochemical ozone creation potential per 1 m^2 plate compared to 5083 alloy.

Compared to previously done studies where environmental footprints were calculated for aluminium ingot mix (European average production and German production) as well as aluminium sheets production of specific compositions: $AlCu_4Mg$, $AlMg_3Mn$ and $AlMg_3$, we can see that innovative Qc5 alloy showed promising results in terms of fossil abiotic depletion potential, eutrophication potential, acidification potential, photochemical ozone creation potential and global warming potential (see Figure 6).

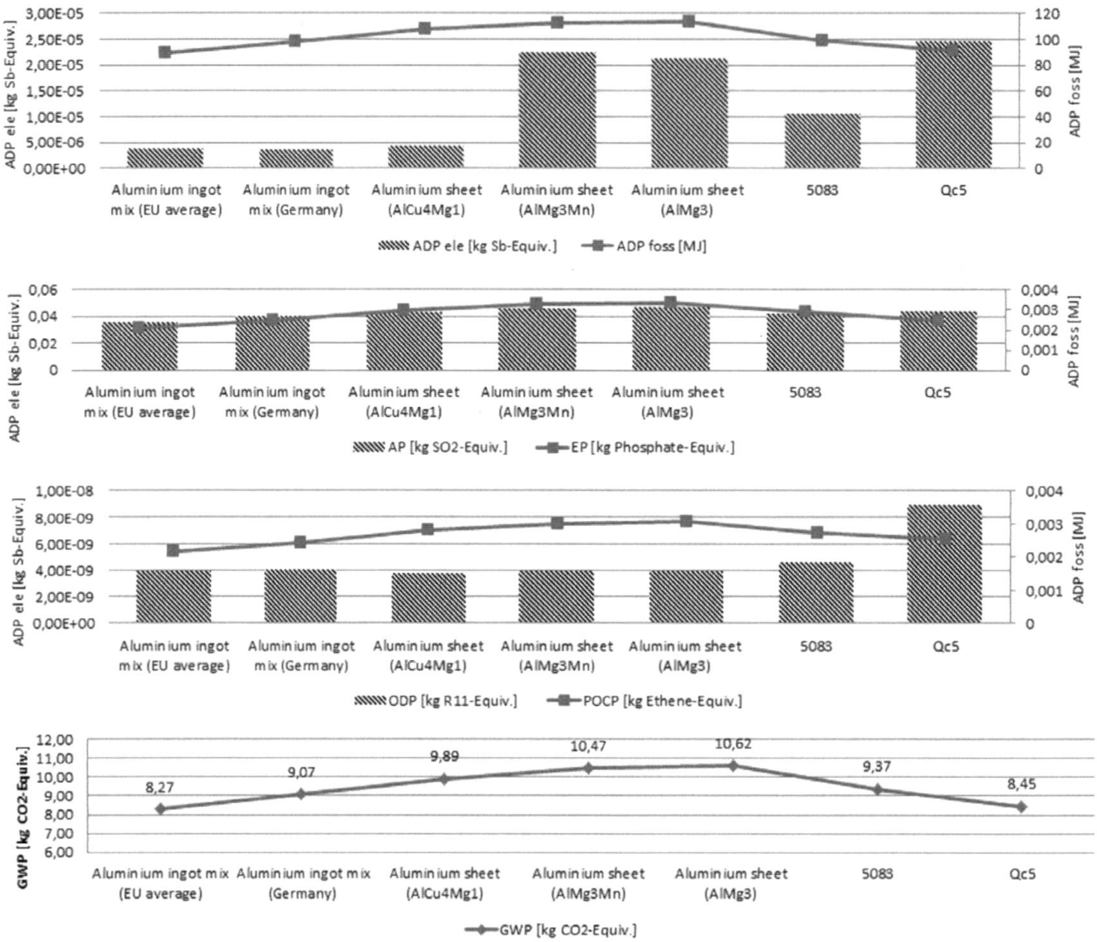

Figure 6. Comparison of environmental footprints for aluminium ingot mix (European average production and German production), sheets production of specific compositions: $AlCu_4Mg$, $AlMg_3Mn$ and $AlMg_3$, 5083 alloy and innovative Qc5 alloy.

A closer look into specific contributors of different environmental footprint parameters is revealed in Figure 7. In terms of GWP, ADP fossile, AP, EP and POCP potentials, aluminium ingot is the biggest contributor by far. Since this is the biggest part also in terms of product mass balance (88 %), the result has been expected. In Qc5 manganese that represents 7.6 % of the mass is the second biggest contributor for those same parameters. ADP elements parameter is mainly influenced by zinc and ODP is caused by zinc and aluminium ingot almost by the equal rate for Qc5 alloy. In 5083 alloy visible contributor in ADP fossile, AP, EP, GWP and POCP is magnesium. Copper plays a major role besides zinc and aluminium ingot for ADO elements.

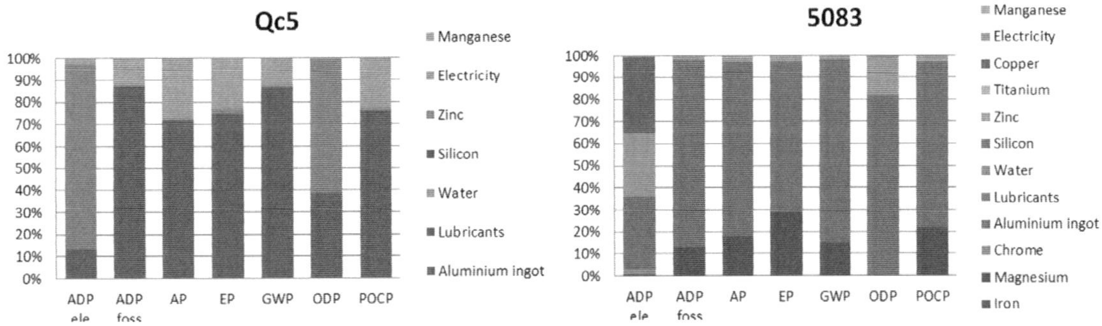

Figure 7. Specific contributors of different environmental footprint parameters for 5083 and Qc5 alloys.

4. Conclusion

An innovative Qc5 alloy was compared to other alloys in terms of environmental footprints related to the production. The study was done according to the standards ISO 14040 and ISO 14044. From cradle to gate (A1-A3 modules) of product's life cycle were studied only. LCA analysis included all input data, such as the basic raw material/materials, auxiliary materials (water) and energy in the production process provided by the producer. Process energy, water and electricity allocation has been provided by researchers proposing an innovative Qc5 alloy and was estimated based on the pilot production of 5083 alloy plates. Qc5 alloy has significantly smaller environmental footprints in terms of global warming potential, eutrophication potential, acidification potential, fossil abiotic depletion potential and photochemical ozone creation potential per 1 m^2 plate compared to 5083 alloy. Compared to previously done studies where environmental footprints were calculated for aluminium ingot mix (European average production and German production) as well as aluminium sheets production of specific compositions: $AlCu_4Mg$, $AlMg_3Mn$ and $AlMg_3$, we can conclude that innovative Qc5 alloy showed promising results in terms of fossil abiotic depletion potential, eutrophication potential, acidification potential, photochemical ozone creation potential and global warming potential.

5. References

[1] ASM Handbook 1992 *Properties and Selection: Nonferrous Alloys and Special-Purpose Materials* (vol 02) ASM International Handbook Committee
[2] Dubois J-M 2012 *Chemical Society Reviews* 41 **20** 6760-77
[3] Dubois J-M 2015 *Useful Quasicrystals* (World Scientific)
[4] Xu D and Han E 2012 *Progress in Natural Science: Materials International* 22 **5** 364-85
[5] Singh A and Tsai A 2008 *Journal of Physics: Condensed Matter* 20 **31** 314002
[6] Naglič I, Delijić K, Samardžija Z, Bikić F, Leskovar B and Markoli B 2015 *Proceedings and Book of Abstracts Belgrade: Association of Metallurgical Engineers of Serbia (AMES)* 67-71
[7] ISO 2006 *Environmental management—life cycle assessment—requirements and guidelines* European Standard EN ISO 14044 International Organization for Standardization
[8] CEN 2012 *Sustainability of construction works—environmental product declarations—core rules for the product category of construction products* European Standard EN 15804 European Committee for Standardization
[9] ISO 2006 *Environmental management—life cycle assessment—principles and framework* European Standard EN ISO 14040 International Organization for Standardization
[10] ASM Handbook 2001 *Composites* (vol 21) ASM International Handbook Committee
[11] Aluminum Association 2013 *The Environmental Footprint of Semi-Finished Aluminum Products in North America* A life cycle assessment report
[12] Nuss P and Eckelman MJ 2014 *PLoS One* **9** 1-12

New materials through a variety of sintering methods

L. Jaworska[1], J. Cyboroń[1], S. Cygan[1], J. Laszkiewicz-Łukasik[1], M. Podsiadło[1], P. Novak[2] and Y. Holovenko[3]

[1] The Institute of Advanced Manufacturing Technology, Wroclawska 37A, 30-011 Krakow, Poland;
[2] Department of Metals and Corrosion Engineering, University of Chemistry and Technology Prague, Technická 5, 16628 Prague, Czech Republic;
[3] Department of Mechanical Engineering, Tallinn University of Technology, Ehitajate tee 5, 19086 Tallinn, Estonia.

E-mail: lucyna.jaworska@ios.krakow.pl

Abstract: New sintering techniques make it possible to obtain materials with special properties that are impossible to obtain by conventional sintering techniques. This issue is especially important for ceramic materials for application under extreme conditions. Following the tendency to limit critical materials in manufacturing processes, the use of W, Si, B, Co, Cr should be limited, also. One of the cheapest and widely available materials is aluminum oxide, which shows differences in phase composition, grain size, hardness, strain and fracture toughness of the same type of powder, sintered via various methods. In this paper the alumina was sintered using the conventional free sintering process, microwave sintering, Spark Plasma Sintering (SPS), high pressure-high temperature method (HP-HT) and High Pressure Spark Plasma Sintering (HP SPS). Phase composition analysis, by X-ray diffraction of the alumina materials sintered using various methods, was carried out. For the conventional sintering method, compacts are composed of α-Al_2O_3 and θ-Al_2O_3. For compacts sintered using SPS, microwave and HP-HT methods, χ-Al_2O_3 and γ-Al_2O_3 phases were additionally present. Mechanical and physical properties of the obtained materials were compared between the methods of sintering. On the basis of images from scanning electron microscope quantitative analysis was performed to determine the degree of grain growth of alumina after sintering.

Keywords: Sintering, SPS, HP SPS, alumina, phases composition, size of grains.

1. Introduction

Research regarding ceramic materials are mainly focused on improving fracture toughness K_{Ic}, by limiting grain size in compacts, improving properties at high temperatures, improving resistance to thermal shock. One of the main goals of modern ceramic technologies is still the manufacture of dense ceramic parts with submicrometric or nanostructured grains. The most important factors which have influence on ceramics properties are:

- powders – their method of obtaining and characteristics (particle size, shape and structure),
- green bodies forming – compression ratio, dimensional changes during compaction,

- sintering – temperature, time and the furnace atmosphere.

There are a lot of material synthesis methods which are helpful in developing new kinds of materials and in powders compaction. The main problem of this, is difficulty with nanopowders compaction, because of large surface area of these powders and huge amount of gases absorbed on the powders surface. There are known techniques of the green bodies ceramic forming, such as: cold pressing, uniaxial pressing, cold isostatic pressing, aqueous injection moulding, direct coagulation casting, electrophoretic forming, gelcasting, hydrolysis assisted solidification, pressure filtration, explosive compaction, lithography [1,2]. The last step for ceramic materials preparation is the sintering process. Sintering of ceramic or metallic powders is a phenomenon that occurs spontaneously at elevated temperatures (below melting point), as a result of the system's desire to lower its surface energy, which decreases as the grains are joined [3]. The goal of the sintering process is to convert highly porous compacted powder into high strength bodies. The process is often accompanied by shrinkage.

The processing of materials is another key parameter that will have implications on the final properties of the ceramic material. Many researchers have observed that the mechanical strength and fracture toughness of ceramics can be improved by the use of different sintering techniques [2]. Variety of sintering methods are available for sintering the ceramic compacts i.e., conventional (free) sintering, reaction-sintering, hot pressing (HP), post-reaction sintering, recrystallization sintering, ultra-high-pressure sintering and isostatic hot pressing (HIP). Generally, the pressure value has strong influence on the sintering. Fig. 1 presents various methods of sintering depending on the type of furnace.

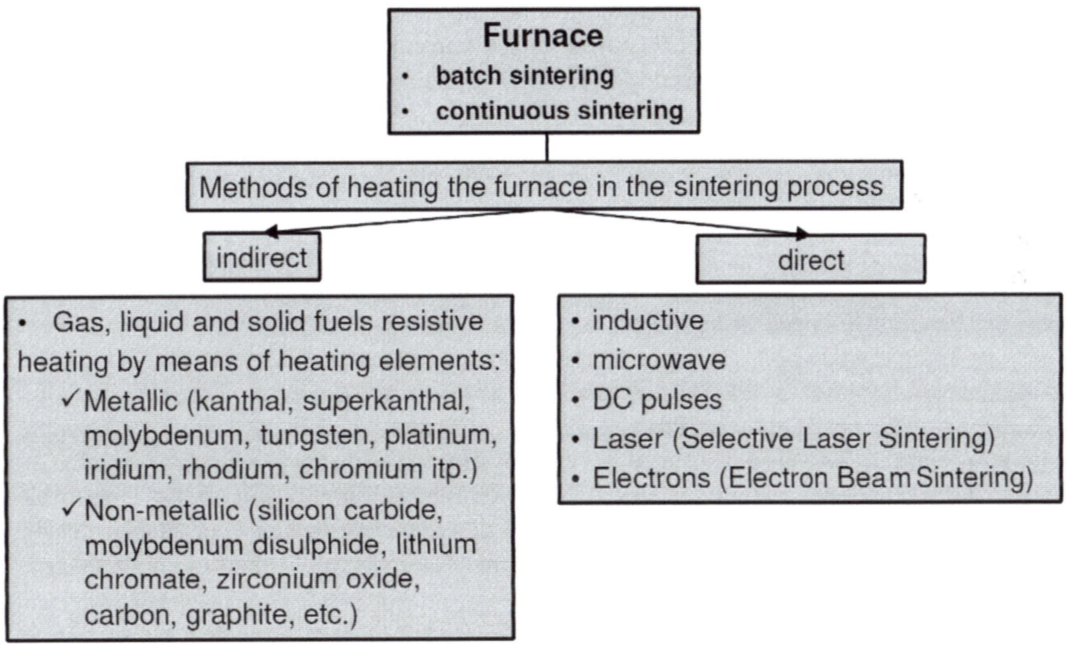

Fig. 1. Different type of furnace using in sintering ceramic composites.

The different physical phenomena used in sintering technologies change thermodynamic conditions of the process, which may affect the crystalline form of the sintered phase and thereby change the properties of the sintered material.

The most characteristic methods of sintering because of physical phenomena are microwave, SPS, HP-HT and SLS methods.

Microwave method has been known since the 1950s [4]. More recent developments in ceramic composites resulted in a new set of challenges for the theory of sintering, particularly when different parts of a structure densify at different rates and temperatures [4]. Many ceramic materials that are difficult to heat at room temperature, possess electrical conductivity or dielectric loss factors that rapidly increase in magnitude as the temperature rises. Thus, these materials will absorb microwave energy if they can be preheated to a suitable temperature using another heat source [4,5]. Microwave sintering process depends on:

- microwave absorption characteristics (dielectric loss factor - low-loss, electrical conductivity),
- size and shape of green bodies,
- sintered materials distribution in the furnace.

Ceramic materials will absorb microwave energy when they can be preheated to suitable temperature using another heat source for example gas or electric furnace or external susceptor material (as SiC) [6]. This kind microwave hybrid heating (MHH) could result in samples with no significant density gradient throughout the cross-section [5].

In the SPS process, a pulsed DC is applied repeatedly from the beginning to the end of the sintering cycle [7]. In the presence of pressure and electric current, localized necking occurs faster, which speeds the sintering process due to joule heating. The temperature rises very fast and the densification is completed within few minutes. The grain boundary area for this method of sintering shows direct grain-to-grain contact, which is attributed to the physical activation of powder particle surface during pulsed current application what has influence on enhanced grain boundary diffusion process [8–10].Three mechanisms may contribute to Spark Plasma Sintering:

- activation of powder particles by pulsed current,
- resistance sintering,
- pressure application [8].

Benefits for the SPS sintering are: reduced sintering duration, good grain to grain bonding, clean grain boundaries, lower temperature of sintering and possibility of sintering materials with significant difference of melting point.

In high pressure high temperature (HP-HT) sintering, special apparatuses is used. The pressure depends on the anvils construction and on the size of the samples. Pressure during the sintering process depends on the apparatus being used. For piston units, it is possible to receive pressure up to 4 GPa, for the Belt type anvils 6-10 GPa, for the Bridgman, the Toroid and the Paris-Edinburg type anvils up to 20GPa. In the multianvils chamber it is possible to generate the pressure in the range 30-100 GPa. Sintered materials should be placed in special gaskets. Limitation for this method is the size of the sintered samples and the high cost of the apparatus assembly parts.

Research concerning the influence of pulsed current employed for heating during high pressure sintering yields very interesting results. For example the results shown in [11] indicate that for HP SPS (High Pressure Spark Plasma Sintering) in comparison to HP-HT technique are characterized lesser amount of phases within the material after sintering. Additionally for HP SPS during the sintering of diamond with Ti and B it was shown that the kinetics of the process results in lower graphite formation, which is very beneficial for this type of materials.

Direct Selective Laser Sintering still does not guarantee high density of most of sintered ceramics powders. The main consolidation mechanism of this process seemed to be partial melting. However, since several mechanisms (partial melting, full melting and solid-state sintering) might act together, it is sometimes not clear which consolidation/binding mechanism is active. The relative density of the alumina compacts prepared using this method was about 85% [11].

Advanced ceramics based on Al_2O_3 exhibit excellent properties such as high thermal resistance, good chemical stability and moderate to high mechanical strength. However, the fracture toughness of the materials is low. The brittleness and poor damage tolerance have so far limited the application of these composites. In relation with that, many authors studied the influenced of additives such as: TiB_2, cBN, Mo or graphene on main properties of the composites [12–15].

The aim of presented work was to investigate the influence of various sintering techniques on microstructure and selected properties of alumina compacts. The alumina was sintered using the conventional free sintering process, microwave sintering, SPS, High Pressure-High Temperature method (HP-HT) and High Pressure Spark Plasma Sintering (HP SPS).

2. Materials and methods

Al_2O_3 (ALCOA, CT300-SG), with an average particle size of 0.7 μm, and 0.3 % MgO (FLUKA), with a particle size in the range of 0.5–1 μm, powders were used. Mixture of powders was prepared in colloidal ball mill through 30 hours with presence of isopropyl alcohol.

For free-sintering and microwave sintering samples were formed by uniaxial pressing in the steel matrix at pressure of 130 MPa and next in the hydraulic press at pressure of 250 MPa.

Green bodies were free sintered during 60 min in the HT 16/18 Nabertherm electric furnace.

For microwave sintering 2.45 GHz in MKH-4.8 Linn High Therm GmbH microwave furnace was applied. Four magnetrons were used which were operating at 50-60% of their full power. The sintered sample was placed in the chamber which were made of porous alumina, and additionally surrounded by SiC susceptors to improve heating.

SPS sintering of starting powder mixtures of alumina powder, without the addition of a temporary binding agent, were placed in a graphite matrix with a diameter 25 mm and pressed at 35 MPa in vacuum. SPS sintering processes were conducted in protective, argon atmosphere during 15 min. Additionally second batch of samples was sintered using SPS method, where the only difference was that, the samples were insulated from the graphite using hBN layer.

HP-HT and HP SPS processes were realized at 4 GPa during 60s, where for HP SPS direct pulsed current was used with 40 ms impulse lenght and 20 ms impulse interval.

Difference between duration of alumina sintering processes are presented in Fig.2.

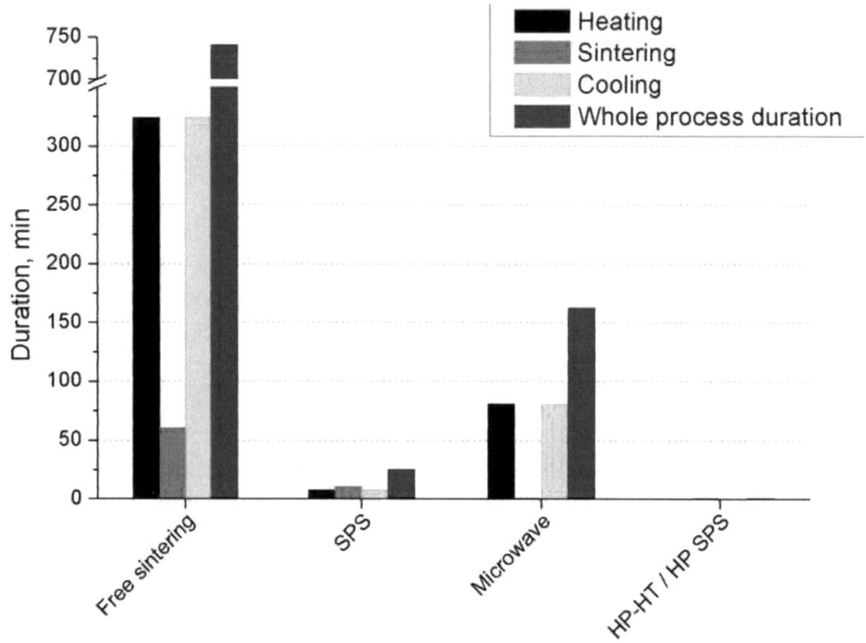

Fig. 2. Comparison of heating, sintering, cooling durations for different type of sintering.

The microstructure investigations were performed using a Jeol JSM-6460LV scanning electron microscope. Density was measured using the hydrostatic method. The hardness was determined by the Vickers method under the load of 9.81N, using a Future Tech FLC-50VX Vickers hardness tester. For each sample, 5 indentations were realized. Young's modulus of the composites was measured with using a Panametrics Epoch III ultrasonic flaw detector; the measurements were based on the transition velocity of the ultrasonic waves through the sample. The velocities of transversal and longitudinal waves were determined as the ratio of the sample thickness and the relevant transition time. Considering that the error in the thickness measurements was ±0.01 mm and in the time-of-flight measurements (±1 ns), the resulting error in the ultrasonic velocity was about 1%. Consequently, the error of Young's modulus could be estimated to be 2%. The X-ray diffraction patterns were obtained by using a PANalitycal Empyrean diffractometer with copper radiation (λCu Kα = 1.5406 Å). The phase analysis was conducted using ICDD (PDF-4+ 2016) files.

The grains size were calculated using the intercept method. According to this method, the average length of grains in the pictures scale of the microstructure was converted to the real given in the μm [16].

3. Results and discussion

Each of sintering method was optimized within the temperature range from 1500°C up to 1650 °C to obtain material with possibly highest values of density and Young's modulus. Table 1 shows the optimized sintering temperatures for microwave, free, SPS, HP-HT and HP SPS sintering processes.

Table 1. Optimized sintering temperature for each of method

Sintering method	Optimized sintering temperature [°C]
Free sintering	1620
Microwave sintering	1620
SPS and SPS with hBN insulation	1550
HP-HT and HP-SPS	1600

The alumina has got many polymorphic varieties. Some of their crystalline structures and physical properties are presented in the Table 2.

Table 2. Main crystallographic structures of Al_2O_3 [17]

Type of Al2O3	Crystal system	Unit cell parameters			Density [g/cm^3]	Temperature of melting point [°C]
		a [nm]	b [nm]	c [nm]		
α-Al_2O_3	Rhombohedral	0.476	0.476	1.299	3.99	2072
γ-Al_2O_3	Cubic	0.794	0.794	0.794	3.47	
δ-Al_2O_3	Tetragonal	0.796	0.796	1.170	3.20	
θ-Al_2O_3	Monoclinic	1.174	0.572	1.124	3.56	
χ-Al_2O_3	Hexagonal	0.563	0.563	2.260	3.35	

The comparison of the alumina phase composition for different sintering method are presented in Fig.3. The diffractogram for the free sintering and SPS sintering with hBN insulation were similar as HP SPS. They are not included in the Fig. 3. Evaluation of the quantitative phase composition of these materials are difficult. Based on the height and width of the peaks coming from θ and χ- Al_2O_3 phase, it can be stated that their total content in the material is about 1.5 - 2% wt. Quantity of the others phases (γ- Al_2O_3 and Al_2O_3- Ref. Nr.-00-012-0539) are at the trace level.

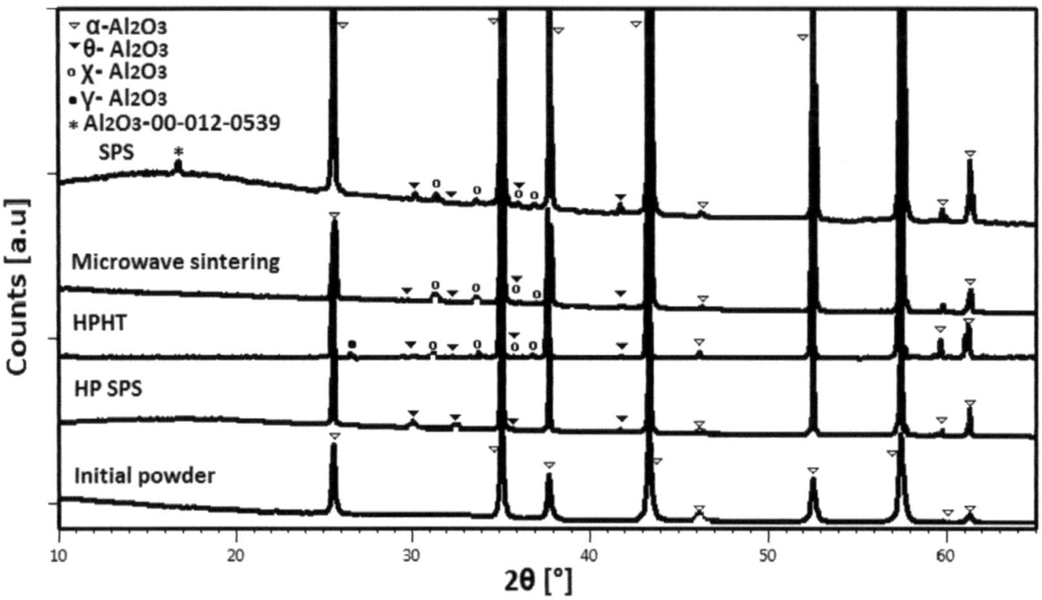

Fig. 3. Phase composition of alumina compacts obtained after different method of sintering.

The comparison of the phase composition for alumina compacts after sintering by various method is presented in the Table 3.

Table 3. Phase composition of alumina compacts obtained after different method of sintering

Sintering method	α-Al_2O_3 R-3m	θ-Al_2O_3 C2m	χ-Al_2O_3 P63/mmc	γ-Al_2O_3 Fd-3m	Al_2O_3 00-012-0539
Initial powder	✓				
Free sintering	✓	✓			
Microwave sintering	✓	✓	✓		
SPS	✓	✓	✓		✓
SPS with hBN insulation	✓	✓			
HPHT	✓	✓	✓	✓	
HP SPS	✓	✓			

Materials after sintering, polishing and etching were subjected to observation under a scanning electron microscope at a magnification 5000x. Microstructures are presented in Fig.4a - Fig.4d. The microstructures and properties of alumina compacts obtained by using High Pressure–High Temperature method were not presented because cracks in the materials. The pressure in the samples during sintering is higher than compressive strength of alumina, so in the compacts severe cracks were present.

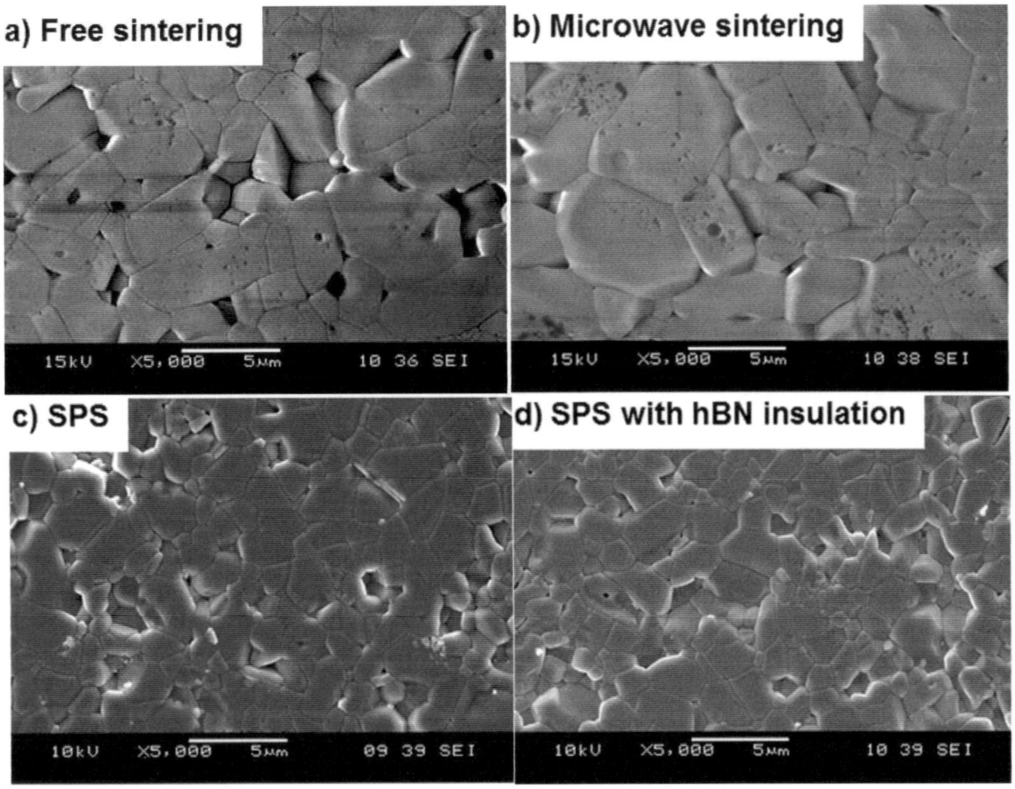

Fig. 4. Microstructures of alumina obtained after: a) free sintering method, b) microwave method, c) SPS method, d) SPS method with hBN insulation.

The use of non-conventional sintering methods, such as Spark Plasma Sintering and microwave sintering, contributes to obtaining a well-densified material with good physical-mechanical properties at relatively short sintering times.

Applied sintering methods allow to receive practically non-porous microstructures of obtained samples with mainly isometric grains of Al_2O_3 (Fig.4a-d).

The size of grains were calculated using the intercept method. In the Table 4 the size of grains for each methods are presented.

Table 4. Grain size of alumina compacts obtained after different method of sintering

Sintering method	**Mean grain size, d [μm]**
Free sintering	2.57±0.12
Microwave sintering	2.86±0.14
SPS	1.41±0.07
SPS with hBN insulation	1.23±0.05

Microwave and free sintering methods did not prevent grains growth of Al_2O_3. Mean grain size of Al_2O_3 is (~2.86 μm) for microwave sintering and (~2.57 μm) for free sintering. The finest grain of alumina was obtained for the SPS sintering.

The results of density, elastic modulus hardness and K_{IC} obtained for all alumina materials are summarized in Table 5.

Table 5. Selected physical and mechanical properties of alumina compacts obtained by various method of sintering

Sintering method	**Density [g/cm3]**	**Young's modulus [GPa]**	**Hardness HV1**	**K_{IC} [MPa*m$^{1/2}$]**
Free sintering	3.91 ± 0.01	383 ± 3	1868 ± 25	5.29 ± 0.31
Microwave sintering	3.87 ±0.01	387 ± 6	1600 ± 106	5.32 ± 0.42
SPS	3.95 ± 0.01	389 ± 4	1806 ± 34	6.80 ± 0.32
SPS with hBN insulation	3.95 ± 0.01	390 ± 5	1803 ± 37	7.17 ± 0.36

The density increases from 3.87 g/cm^3 up to 3.95 g/cm^3 for materials obtained after microwave and SPS sintering, respectively. It is increasing with alumina χ-Al_2O_3 content in the material is fallen. This is related to the theoretical density of χ-Al_2O_3 which is the smallest for each of forms of alumina. In the course of these studies, no significant differences were found in hardness values for Al_2O_3 materials after various methods of sintering, apart from the results for material sintered by microwave method. The smallest hardness for this material could be connected with additional form of alumina (χ-Al_2O_3). In the remaining materials, if there is χ-Al_2O_3 present, its properties are compensated by other forms which are characterized by higher density and different crystalline structure.

4. Conclusion

- The phase composition of the alumina compacts after SPS sintering, SPS with hBN insulation and HP SPS is similar, composed of α-Al_2O_3 and θ-Al_2O_3, for HP-HT, SPS (without insulation) and microwave sintering the χ-Al_2O_3 appears in the material.
- The χ-Al_2O_3 participation has influence on density and hardness of compacts.
- High pressure methods above 4 GPa generated severe cracks in the material.
- Materials obtained using the SPS methods are characterized by better compaction and higher mechanical properties, size of grains is lower than for materials obtained using microwave and free sintering methods.

Acknowledgments

This research was supported by COST Action CA15102 – "Solutions for Critical Raw Materials Under Extreme Conditions" (CRM-EXTREME) and by the statutory activity grant to the Institute of Advanced Manufacturing Technology, Cracow, Poland (DS-17.3.2).

References:

[1] Sigmund W M, Bell N S and Bergström L 2004 Novel Powder-Processing Methods for Advanced Ceramics *Journal of the American Ceramic Society* **83** 1557–74
[2] Deckers J, Vleugels J and Kruth J P 2014 Additive manufacturing of ceramics: A review *Journal of Ceramic Science and Technology* **5** 245–60
[3] A.Lakshmanan E by 2012 Sintering of Ceramics - New Emerging Techniques
[4] Von Hippel A R 1954 Dielectric Materials and Applications
[5] Menezes R R, Souto P M and Kiminami R H G A 2012 Microwave Fast sintering of ceramic materials *Sintering of ceramics - New Emerging Techniques* (InTech)

[6] Krage M 1981 Microwave sintering of ferrites *American Ceramic Society Bulletin,* **60** 1232–4
[7] Suárez., Fernández J, L. Menéndez R, Torrecillas H, U. Kessel J, Hennicke R and Kessel K and T 2013 Challenges and Opportunities for Spark Plasma Sintering : A Key Technology for a New Generation of Materials *Sintering Applications* (InTech)
[8] Munir Z A, Anselmi-Tamburini U and Ohyanagi M 2006 The effect of electric field and pressure on the synthesis and consolidation of materials: A review of the spark plasma sintering method *Journal of Materials Science* **41** 763–77
[9] Tokita M 2013 Spark Plasma Sintering (SPS) Method, Systems, and Applications *Handbook of Advanced Ceramics: Materials, Applications, Processing, and Properties: Second Edition* 1149–77
[10] Kessel H U 2009 Sintered Materials on the Way to Production by Means of Modern SPS Technologies *Ceramic Forum International Berichte Der Deutschen Keramischen Gesellschaft* **10** 145–52
[11] Deckers J, Meyers M, Wijremblewski B, Vleugels J and Kruth J P 2014 Direct selective laser sintering/melting of fine grained alumina with a novel experimental setup *Mater. Sci. Tech.: Special issue on 3D printing*
[12] Putyra P, Jaworska L, Podsiadło M, Laszkiewicz-Łukasik J and Krzywda T 2016 The ceramic materials to forming by EDM method *ISNNM-2016 The 14th International Symposium on Novel and Nano Materials*
[13] Klimczyk P, Cura M E, Vlaicu A M, Mercioniu I, Wyzga P, Jaworska L and Hannula S P 2015 Al_2O_3-cBN composites sintered by SPS and HPHT methods *Journal of the European Ceramic Society* **36** 1783–9
[14] Broniszewski K, Wozniak J, Kostecki M, Czechowski K, Jaworska L and Olszyna A 2015 Al_2O_3–Mo cutting tools for machining hardened stainless steel *Wear* **41** 14190–6
[15] Morgiel J, Klimczyk P, Major Ł, Jaworska L, Maj Ł, Cygan T and Olszyna A 2017 TEM investigations of wear mechanism of Al_2O_3 and Si_3N_4 compacts with GLPs additions *Ceramics International* **43** 8334–42
[16] Ryś J 1995 *Stereologia materiałów* (Kraków: FotoBit Design)
[17] Programme 2016 ICDD base PDF 4+2016

IOP Conf. Series: Materials Science and Engineering **329** (2018) 012005 doi:10.1088/1757-899X/329/1/012005

Possible alternatives to critical elements in coatings for extreme applications

Maria Luisa Grilli[*1], Daniele Valerini[2], Radu Robert Piticescu[3], Tiziano Bellezze[4], Mehmet Yilmaz[5], Antonio Rinaldi[1], Santiago Cuesta-López[6,7], Antonella Rizzo[2]

[1]ENEA - Italian National Agency for New Technologies, Energy and Sustainable Economic Development, Casaccia Research Centre, Via Anguillarese 301, 00123 Roma, Italy

[2]ENEA - Italian National Agency for New Technologies, Energy and Sustainable Economic Development, Brindisi Research Centre, S.S. 7 Appia - km 706, 72100 Brindisi, Italy

[3]National R&D Institute for Nonferrous and Rare Metals, Biruintei Blvd. No. 102, 077145 Pantelimon, Romania

[4]Department of Materials, Environmental Sciences and Urban Planning (SIMAU), Polytechnic University of Marche, Via Brecce Bianche 12, 60131 Ancona, Italy

[5]Department of Science Teaching, Faculty of K. K. Education, Atatürk University, 25240 Erzurum, Turkey

[6] ICCRAM - Int. Center for Critical Raw Materials and Advanced Industrial Technologies Plz. Misael Banuelos,09001 Burgos, Spain

[7]ICAMCyL Foundation. International Center for Advanced Materials and Raw Materials of Castilla y Leon, 24492 Cubillos del Sil (León), Spain

*E-mail: marialuisa.grilli@enea.it

Abstract. Surface functionalisation and protection have been used since a long time for improving specific properties of materials such as lubrication, water repellence, brightness, and for increasing durability of objects and tools. Among the different kinds of surface treatments used to achieve the required properties, the use of coatings is fundamental to guarantee substrate durability in harsh environments. Extreme working conditions of temperature, pressure, irradiation, wear and corrosion occur in several applications, thus very often requiring bulk material protection by means of coatings. In this study, three main classes of coatings used in extreme conditions are considered: i) hard and superhard coatings for application in machining tools, ii) coatings for high temperatures (thermal barrier coatings), and iii) coatings against corrosion. The presence of critical elements in such coatings (Cr, Y, W, Co, etc.) is analysed and the possibility to use CRMs-free substitutes is reviewed. The role of multilayers and nanocomposites in tailoring coating performances is also discussed for thermal barrier and superhard coatings.

1. Introduction

Profound understanding of the role of critical elements in the properties of high tech bulk and coating materials is a prerequisite to design new materials with reduced or no critical elements content. Substitution of critical raw materials (CRMs) is one of the challenges of the European Innovation Partnership (EIP) on Raw Materials, together with the improvement of the mining processes and the increase in recovery and recycling rates. The criticality of a raw material for the EU is defined basing on its supply risk and economic importance for the EU, which depend on geopolitical issues and can therefore change in a dynamic way. For this reason CRMs list is updated periodically by the EC, which released the latest list of CRMs in September 2017 [1].

Table 1. List of critical raw materials (CRMs) for the EU [1].

CRMs (as defined in 2017 by EC)			
Antimony	Gallium	Magnesium	Scandium
Barite	Germanium	Natural graphite	Silicon metal
Beryllium	Hafnium	Natural rubber	Tantalum
Bismuth	Helium	Niobium	Tungsten
Borate	HREEs	PGMs	Vanadium
Cobalt	Indium	Phosphate rock	
Fluorspar	LREEs	Phosphorous	

With the aim to reduce EU dependence on these raw materials, both substitution or reduction in the use of CRMs should be pursued, although even reduction represents often a challenging task. This goal becomes even more important when the considered CRM has got a high impact on environment and health. We have already examined the substitution of CRMs in bulk materials in a previous study [2], where we highlighted also the fundamental role of coatings for protection of bulk materials for application under extreme conditions in terms of wear, temperature and corrosion, and for saving critical elements. Literature is very reach of examples of high performing single layer and multilayer coatings, however, none of the papers reports about the role of critical elements in such coatings. In this study, we review on hard, thermal barrier and anticorrosion coatings, analysing CRMs content (Cr, Y, W, Co, etc.) and describing already existing or novel CRMs-free solutions.

2. State of the art of coatings for extreme conditions and analysis of CRMs content

2.1 Hard coatings for machining tools

With the increasing of cutting tools speed, the demand for heat resistant materials has increased, resulting in a progressive substitution of high-speed steels with carbides and superhard materials as main tool constituents. The cemented carbides are a set of composite materials that consist of hard carbide particles in a metallic matrix. The most commonly used material is WC-Co. It is estimated that more than 60% of cemented carbides tools are coated to increase their lifetime and to allow faster machining. Examples of coatings for cutting tools include nitrides (CrN, TiN, Ti(C)N, TiAlN, ZrN, ZrAlN, a-CN), diamond-like carbon (DLC), oxides (Al_2O_3, TiO_2, etc.) and carbides (SiC, WC, VC, TiC, Cr_3C_2, BC, etc.). Many of these coatings are already CRM-free, so the quest for further optimized coating composition and architecture becomes important to save the CRMs contained in the underlying bulk tools (e.g. W and Co). Apart from economic and supply reasons, another driving force for Co substitution is its toxicity, since the WC-Co mixture is considered as a possible carcinogenic material, according to the International Agency for Research on Cancer.

Among the different coating materials used on machining tools, the most widespread are represented by DLC and nitride-based coatings like monolayers and multilayers based on AlTiN, AlCrN, TiCN, as well as quaternary coatings, which are grown by Physical Vapour Deposition (PVD) or Plasma Assisted Chemical Vapour Deposition (PACVD) techniques. They currently represent one of the best solutions due to their high thermal stability, hot hardness and oxidation resistance, and they can be used to coat tools made of different materials (steel alloys, cemented carbides, Ti alloys, etc.). In TiAlN coatings, alloying with Al improves the hardness and the oxidation resistance and the thermal stability of the TiN material, thanks to a dense Al_2O_3 scale forming on the surface of the coating [3]. Other alloying elements are Cr, Y, Zr, V, Si, B and Ta, where: i) Cr and Y are known for improving oxidation and corrosion resistance [4-6]; ii) Zr improves the wear resistance of TiN based coatings, stabilizes the fcc TiN lattice and forms a very thin stable oxide layer similar to Al_2O_3 [3,4]; iii) V increases TiN hardness [4,7]; iv) Si increases the hardness and chemical resistance of (Ti,Al)N coatings [8]; v) B in (Ti,Al)N lattice improves the lubricated cutting performance [4]; vi) Ta improves the hardness, thermal stability and oxidation resistance, thus increasing maximum milling time with respect to conventional TiAlN [9].

High Velocity Oxygen Fuel (HVOF) sprayed WC-Co and WC-Co-Cr coatings deposited on high strength steel (HSS) are also considered as an alternative to the use of bulk WC-Co [10].

2.2 Thermal barrier coatings

Thermal barrier coatings (TBCs) are high refractory materials applied to metallic (generally Ni superalloys) surfaces of components such as gas turbine or aero-engine parts, which operate at elevated temperatures. These coatings are used to insulate structural components from large and prolonged heat loads, limiting the thermal exposure and therefore extending their durability by reducing oxidation and thermal fatigue. The insulating capability of TBCs enables higher operating temperatures and/or permits a reduction in the required amount of cooling air, thereby improving efficiency, reducing emission and increasing thrust/weight ratio. A typical TBC system is composed of a metallic bond coat adherent to the metallic substrate and a ceramic top coat (Fig. 2.). In these systems the main role of the superalloy substrate is to support the mechanical load. The role of the very thin bond coat (mainly NiCoCrAlY or NiAl based alloys) is to protect the metal substrate from oxidation/corrosion, particularly from oxygen and corrosive elements that penetrate the porous ceramic top coat. Bond coat also contains the source of elements to create the next thermally grown oxide (TGO) layer that provides bonding of TBC to bond coat and slows subsequent oxidation.

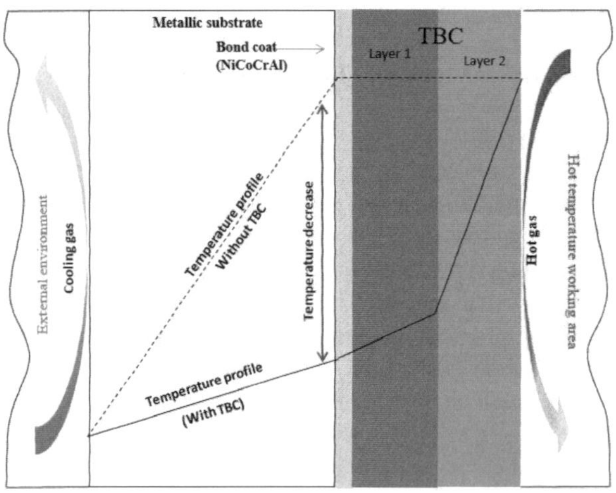

Figure 1. Schematic representation of ceramic thermal barrier coatings.

Critical elements are contained both in the bond and top coat. The most commonly used top coat is Yttria Stabilized Zirconia (YZS), i.e. ZrO_2 partially stabilised with about 6–8 wt % (3-4 to 6-8 mol %) Y_2O_3, usually deposited by Electron Beam Physical Vapour Deposition (EB-PVD) and Plasma Spray techniques. Alumina, mullite, ceria and perovskites are also used due to their high melting point and low thermal conductivity [11]. Commercial zirconia powders doped with 9.5%Y_2O_3-5.6%Yb_2O_3-5.2%Gd_2O_3 Metco 206A agglomerated powder have been proposed for fabrication of thermal barrier coatings operating at temperatures up to 1500°C and they are considered as one of the best available material system for thermal protection of propulsion systems for aeronautics [12].
State of the art of bond coat is based on MCrAlY ternary system, where M stands for Fe, Co or Ni [13].

2.3 Coatings against corrosion
According to National Association of Corrosion Engineers (NACE), worldwide cost of corrosion has been estimated to be nearly $2,5 trillion per year [14]. Prevention is the best practice to fight against corrosion. Coatings are essential for corrosion protection because they provide a physical barrier which impedes the access of aggressive species to the metallic substrates and/or inhibit the corrosion process. The chemical species largely used for long time against corrosion is Cr(VI), due to its ability to form a passive oxide layer (chromate conversion coating (CCC)) on the metallic substrates (e.g. Zn, Al, Mg based alloys). These layers not only provide an effective physical barrier against corrosive agents, but they have also the self-healing ability in the event of surface damage [15]. Despite its high economic importance, Cr was recently removed from the latest CRMs list of the EC, due to the decrease of its supply risk. However it is right at the edge of the criticality threshold for the EU, and it remains still a high value metal with a fundamental role in protection against oxidation. In addition, due to its toxicity, REACH (Registration, Evaluation, Authorisation and Restriction of Chemicals) identified Cr(VI) among substances of very high concern (SVHC), the use of which is restricted and must be authorised [16]. Cr(VI) use is actually allowed in aeronautic and aerospace industries, even though great concern is also increasing in these sectors.

Cr is contained in several coatings: i) chromate conversion coatings and corrosion inhibitors in pigments for preventing wet/aqueous corrosion; ii) NiCrAlY materials for protection against oxidation in high temperature gas environments; iii) CoCrAlY for improving the resistance to sulfidation; iv) NiCoCrAlY to provide an acceptable balance of both oxidation and sulfidation resistance; v) thermal-sprayed WC-CrNi, Cr_3C_2–NiCr, WC-CoCr, etc. for protection against corrosion, erosion, and wear.

3. Beyond the state of the art: CRMs-free solutions

Concerning the coatings for cutting tools, one of the best choices is the use of CRM-free nitride-based coatings like (Ti,Al)N, TiCN, etc., deposited by PVD. Indeed, thanks to their good balance of hardness, wear resistance and high-temperature oxidation resistance, they represent an optimal improvement for protection of machining tools and mechanical components. Further enhancements in the properties of hard PVD coatings can be achieved by using graded layers, nanocomposite layers, multilayers or superlattices [17,18,19]. For example, nanocomposite superhard coatings (H>40 GPa), formed by nanometer size MeN particles (Me = metal element) embedded in amorphous or crystalline matrices, the so called "third generation ceramic coatings", exhibit enhanced mechanical, electronic, magnetic and optical properties, thanks to their nanocrystalline structure made by nanometric (5-10 nm) features.

Superlattices are multilayers obtained by alternating several layers of different materials with nanometric thickness (up to 10 nm). An example of ZrN/TiN superlattice deposited by PVD is reported in Fig. 3. Similar multilayers were used to coat a WC-Co tool, resulting in an increased milling time of about 30% [20]. Due to their enhanced mechanical properties, such coatings may be applied to CRM-free substrates having lower performances with respect to WC-Co, so to tailor those to ones necessary for cutting tools, at least in some applications. Possible CRM-free bulk materials candidates are composites having NiAl matrix and ceramic reinforcement (Al_2O_3), that were recently developed by reactive sintering and successfully tested. Even though the wear resistance does not reach the performances of WC-Co composites, it exceeds the properties of current cold work tool steels without the need for heat treatment [21]. At present, these materials are relatively brittle at room temperature and therefore need further optimization, in terms of chemical composition and/or production technology, in order to achieve the necessary ductility for the application as tool materials [2].

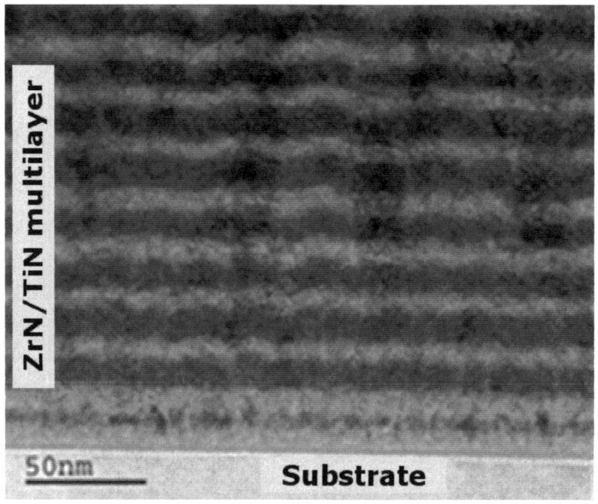

Figure 2. Cross-sectional high-resolution TEM image of a ZrN/TiN multilayer [20].

Besides the proper choice of coating composition and structure, alternative solutions to get coatings with further enhanced protective characteristics may be found in the use of innovative deposition techniques, such as the HiPIMS (High Power Impulse Magnetron Sputtering) [22, 23]. This PVD technique is an evolution of the conventional sputtering technique that exploits short pulses with a very high power, giving rise to a very high plasma density and a high ionization fraction and energy of the sputtered species, together with a high rate of dissociation of the gaseous species in the deposition chamber. Compared to other standard deposition techniques like cathodic arc and conventional magnetron sputtering, the HiPIMS techniques has been reported to allow the deposition of hard coatings presenting higher density, lower content of defects and droplets, and better adhesion to the coated component. Thanks to these improved properties, nowadays HiPIMS-deposited coatings are widely available as commercial protective coatings.

Nanostructured coatings, superlattices and functionally graded coatings are also very important to tailor properties of thermal barrier coatings and reduce mismatch and thermal gradients between top coat and bond coat, which are among the reasons of coatings' failure [24-26]. In addition, they can provide a better barrier for oxygen diffusion.

The possible alternatives to YSZ top coats are constituted by Mullite ($3Al_2O_3 \cdot 2SiO_2$), $La_2Zr_2O_7$, $BaZrO_3$, α-Al_2O_3, ZrO_2, Al_2O_3-TiO_2 and metal- glass composites (MGCs) [27]. Among these alternatives, some are CRM-free, but at present do not show comparable performances at high temperature with respect to YSZ. More recently, the search has been broadened by using insights from atomistic simulations and crystal chemistry, searching for new approaches for reflecting or scattering radiation and preventing it from reaching the metallic parts [28, 29]. Finally, novel compositions of high entropy alloys [30, 31] with reduced CRMs content, both metallic and oxides, could be investigated as possible bond coat and top coat, respectively.

Regarding coatings against corrosion, several alternatives to CCC have been proposed. Organosilane compounds, having the general formula R_nSiX_{4-n} (where X is the hydrolyzable group and R is an organofunctional group), are extensively investigated. These compounds show good corrosion protection properties when are deposited on different types of metal substrates: they form a barrier constituted by Si-O-Si networks [32-35]. Addition of corrosion inhibitors may further increase their corrosion resistance [36]. Highly crosslinked silane polymers made by a bis-sulfur/bis-amino (3/1) mixture showed good corrosion protection properties for many metals including aluminum alloys and galvanized steel [11].

Other already investigated alternatives to Cr are chemical conversion coatings of different nature: (i) phosphate–permanganate, fluoride and stannate for the protection of magnesium alloys [37, 38]; (ii) molybdates for zinc alloy [39] and aluminium alloy substrates [40]; (iii) permanganates, vanadates and Zr- and Ti-based coatings [40]. Some of the listed species were used as primers before the application of a top coat obtained by sol-gel or PVD techniques [41].

Recent findings showed also that compact and defect free TiO_2 sol gel coatings, deposited at low temperature directly (i.e. without the need of a primer) on AA6082 substrate, showed improved corrosion resistance with respect to the uncoated alloy [42].

Almost all authors who studied these solutions agree on the fact that they limitedly represent a viable alternative to CCC, apart from some cases where less demanding corrosion conditions are present.

In a recent work, however, the corrosion resistance of a commercial processes (PreCoat A32) on Al alloys was found comparable or even better than that of the widely used conventional CCC Alodine 1200 [43].

Finally, much emphasis is given to self-healing and smart coatings with tailored functionalities for improved corrosion protection [44, 45].

4. Conclusions

Among the actions aimed to decrease the EU dependence on critical raw materials, it is important to find new solutions capable to replace or reduce the content of CMRs in several technological fields, due to reasons related to economic, supply and toxicity issues, although substitution and even reduction of CRMs content is difficult especially in certain applications. In particular, in this paper the presence of critical elements in coatings against wear, high temperature and corrosion is analysed together with the possibility to reduce their content. Possible alternatives are investigated in terms of use of other CRMs-free coatings, optimization of coating structure (nanocomposite coatings and superlattices), and exploitation of advanced deposition techniques, so to obtain coatings with enhanced performances thus allowing lower CRMs content or at least saving critical elements that are present in the coated components.

Acknowledgments

COST Action "Solutions for Critical Raw Materials under Extreme Conditions", supported by COST (European Cooperation in Science and Technology) and H2020 Grant Agreement TWINNING 692216 "The virtual Center for sustainable development of Advanced Materials operating under extreme conditions"- Acronym SUPERMAT are acknowledged.

References

[1] Study on the review of the list of Critical Raw Materials, September 2017, DOI: 10.2873/876644. Available online: https://publications.europa.eu/en/publication-detail/-/publication/08fdab5f-9766-11e7-b92d-01aa75ed71a1/language-en

[2] Grilli ML, Bellezze T, Gamsjäger E, Rinaldi A, Novak P, Balos S, Piticescu RR and Ruello ML 2017 *Materials*, **10**, 285, doi:10.3390/ma10030285

[3] Knotek O, Bohmer M, Leyendecker T, and Yungblut F 1988 *Mater. Sci. Eng.* **A 105/106**, 481

[4] PalDey S, Deevi SC 2003 *Materials Science and Engineering A*, **342**, 58-79
[5] Jehn HA, Thiergarten F, Ebersbach E and Fabian D 1991 *Surf. Coat. Technol.* **50**, 45
[6] Zhujing J, Changqing L, Li Y and Weitao W 1991 *Surf. Coat. Technol.* **46**, 307
[7] Davies KE, Gan BK, McKenzie DR, Bilek MMM, Taylor MB, McCulloch DG and Latella BA 2004 *J. Phys.: Condens. Matter* **16**, 7947
[8] Takahashi K et al. 2011 *Jpn. J. Appl. Phys.* **50**, 075802
[9] Hollerweger R, Riedl H, Paulitsch J, Arndt M, Rachbauer R, Polcik P, Primig S and Mayrhofer PH 2014 *Surf. Coat. Technol.*, **257**, 78
[10] Tillmann W et al 2017 *IOP Conf. Ser.: Mater. Sci. Eng.* **181** 012011
[11] Cao XQ, Vassen R and Stoever D. 2004 *J. Eur. Ceram. Soc.*, **24**, 1-10
[12] DSMTS-0099.5 – Zirconia Gadolinia Ytterbia Yttria Agglomerated and Sintered Powder, © 2017 Oerlikon Metco, www.oerlikon.com/metco
[13] Li WZ, Wang QM, Bao ZB, Yao Y, Gong J, Sun C and Jiang X 2008 *Materials Science and Engineering A*, **498**, 487–494
[14] Study Sets Course Toward Corrosion Management Practices to Increase Safety, Decrease $2.5 Trillion Global Cost of Corrosion, 2016-03-07. Available online: https://www.nace.org/Newsroom/NACE-News/Study-Sets-Course-Toward-Corrosion-Management-Practices-to-Increase-Safety,-Decrease-$2-5-Trillion-Global-Cost-of-Corrosion/
[15] Zarras P and Stenger-Smith JD 2015 Smart Inorganic and Organic Pretreatment Coatings for the Inhibition of Corrosion on Metals/Alloys *Intelligent Coatings for Corrosion Control*, Eds, A Tiwari, J Rawlins, LH Hihara, B Heinemann, (Elsevier, Oxford UK, Waltham, MA USA) chapter 3 pp 59-91
[16] REACH and the impact of Hexavalent Chromium. Available online: http://www.sea.org.uk/blog/reach-and-the-impact-of-hexavalent-chromium.htm
[17] Musil J 2000 *Surf. Coat. Technol.* **125**, 322
[18] Zhang S, Sun D, Fu Y and Du H 2003 *Surf. Coat. Technol.* **167**, 113
[19] Martinu L, Zabeida O and Klemberg-Sapieha JE 2010 Plasma Enhanced Chemical Vapor Deposition of Functional Coatings *Handbook of Deposition Technologies for Films and Coatings*, Ed. PM Martin, (Elsevier, Columbia Basin Thin Film Solutions LLC, Kennewick, WA, United States) chapter 9 pp 392-465
[20] Rizzo A, Signore MA and Valerini D 2012 *EAI Energia*, **3**, 102-108
[21] Novák P, Šotka D, Novák M, Michalcová A, Šerák J, Vojtěch D 2011 *Powder Metall.*, **54**, 308-313
[22] Sarakinos K, Alami J and Konstantinidis S 2010 *Surf. Coat. Technol.* **204**, 1661
[23] Gudmundsson JT, Brenning N, Lundin D and Helmersson U 2012 *J. Vac. Sci. Technol. A*, **30**, 030801
[24] Clarke DR and Levi CG 2003 *Annual Review of Materials Research*, **33**, 383-417
[25] Di Girolamo G, Blasi C, Brentari A and Schioppa M 2013, *EAI Energia* **1-2**, 69-76
[26] Di Girolamo G and Serra E 2015 Thermally sprayed nanostructured coatings for anti-wear and TBC applications: State-of-the-art and future perspectives, *Anti-Abrasive Nanocoatings Current and Future Applications* (Edition: Elsevier Ltd), pp 513-541
[27] Huibin X and Hongbo G 2011 *Thermal barrier coatings* (Cambridge: Woodhead Publishing Limited)
[28] Clarke DR and Phillpot SR 2005 *Materialstoday*, **8**, 22-29
[29] Clarke DR, Oechsner M and Padture NP 2012 *MRS Bullettin*, **37**, 891-898
[30] Gao MC 2013 *JOM*, Vol. 65, 1749-1750
[31] Murty BS, Yeh JW and Ranganathan S 2014 *High-entropy alloys* (London: Butterworth-Heinemann)
[32] Calabrese L, Bonaccorsi L and Proverbio E 2012 *J. Coat. Technol. Res.*, **9**, 597–607
[33] Zaferani SH, Peikari M, Zaarei D, Danaee I, Fakhraei JM and Mohammadi M 2012 *Corrosion* **69**(4), 372-387
[34] Schaefer DW, Pan G and vanOoij W 2006 *Los Alamos Science*, **30**, 172-177

[35] Brusciotti F, Snihirova DV, Xue H, Montemor MF, Lamaka SV and Ferreira MGS 2013 *Corrosion Science,* **67**, 82-90
[36] Palimi MJ, Rostami M, Mahdavian M and Ramezanzadeh B 2015 J *Coat Technol Res*, **12**, 277
[37] Pommiers S, Frayret J, Castetbon A and Potin-Gautier M 2014 *Corrosion Science*, **84**, 135–146
[38] Chen XB, Birbilis N and Abbott TB 2011 *Corrosion*, **67**(3), 035005-1 – 035005-16
[39] Walker DE and Wilcox GD 2008 *Transactions of the Institute of Metal Finishing*, **86**(5), 251-259
[40] Kulinich SA and Akhtar AS 2012 *Russian Journal of NonFerrous Metals*, **53**(2), 176–203
[41] Bagalà P, Lamastra FR, Kaciulis S, Mezzi A and Montesperelli G 2012 *Surface & Coatings Technology*, **206,** 4855–4863
[42] Mori S, Lamastra FR, Kaciulis S, Soltani P and Montesperelli G 2017 *Corrosion Engineering Science and Technology*, in press
[43] Carreira AF, Pereira A M, Vaz EP, Cabral A M, Ghidini T, Pigliaru L and Rohr T, 2017 *J. Coat. Technol. Res.,* **14** (4) 879–892
[44] Cho SH, White SR, and Braun, PV 2009, *Advanced Materials*, **21**, 645–649
[45] Montemor M F 2014, *Surface & Coatings Technology,* **258**, 17–37

IOP Conf. Series: Materials Science and Engineering **329** (2018) 012006 doi:10.1088/1757-899X/329/1/012006

Mechanical characterization of alloys in extreme conditions of high strain rates and high temperature

Ezio Cadoni

DynaMat Laboratory, University of Applied Sciences of Southern Switzerland, Campus SUPSI-Trevano, 6952 Canobbio, Switzerland

ezio.cadoni@supsi.ch

Abstract. The aim of this paper is the description of the mechanical characterization of alloys under extreme conditions of temperature and loading. In fact, in the frame of the Cost Action CA15102 "Solutions for Critical Raw Materials Under Extreme Conditions (CRM-EXTREME)" this aspect is crucial and many industrial applications have to consider the dynamic response of materials. Indeed, for a reduction and substitution of CRMs in alloys is necessary to design the materials and understand if the new materials behave better or if the substitution or reduction badly affect their performance. For this reason, a deep knowledge of the mechanical behaviour at high strain-rates of considered materials is required. In general, machinery manufacturing industry or transport industry as well as energy industry have important dynamic phenomena that are simultaneously affected by extended strain, high strain-rate, damage and pressure, as well as conspicuous temperature gradients. The experimental results in extreme conditions of high strain rate and high temperature of an austenitic stainless steel as well as a high-chromium tempered martensitic reduced activation steel Eurofer97 are presented.

1. Introduction

The knowledge of mechanical dynamic behaviour of materials under extreme conditions of high strain rates and high temperature allows the development of reliable material models and their use in several industrial sectors such as energy, automotive, transportation and machine manufacturing. The mechanical characterization requires the use of adequate techniques able to describe the behaviour of these materials in such severe conditions of loads and temperature. In fact, the accuracy of the mechanical assessment and the design of products made of high value alloys and metal-matrix composites subjected to rapid loading and high temperature strongly depends on the precision of the material's constitutive law in reproducing the real behaviour of such materials. As a consequence, the mechanical characterization is one of the fruitful ways in order to find viable alternatives to CRMs and promote the industrial exploitation of substituted materials. The optimisation of such alternative materials should then pass through high strain rate testing phase that are obtained by imposing an acting pulse by means of special dynamic testing devices. These apparatuses generate a stress wave pulse well controlled in amplitude and duration which is propagated without dispersion and uncontrolled reflections to load and deform until fracture a specimen and to allow the record analysis by using the well proofed uniaxial elastic waves propagation theory. The methodology proposed for this characterization is the Split Hopkinson-Kolsky Bar technique, that is the scientifically most recognized methodology for the precision measurements of pulses parameters, allow the characterization of the material dynamic properties. This contribution aims to present a precision measurement methodology of the material stress-strain curves carried out by means of a modified Hopkinson bar used in the DynaMat Laboratory. The results at high strain rate with high temperature of austenitic stainless steels as well as high-chromium tempered martensitic reduced activation steel Eurofer97 will be presented.

2. Experimental techniques for high strain rate regimes

An innovative version of the split Hopkinson bar, which can be used for performing tension, compression and shear tests, in the strain-rate range 10^2 to 10^3 s^{-1} (10^4 s^{-1} in shear), was developed and patented at the Joint Research Centre of the European Commission in the seventies in the frame of nuclear reactor safety studies of severe accidents [1-2]. In the last decades, this technology has been extensively adopted in many other fields [3-11].

The development of the JRC-Modified split Hopkinson bar was particularly needed in order to dispose of a unique versatile equipment capable of working in tension, compression and shear, generating the long duration pulses. Using the classic SHPB dynamic testing equipment it would have been necessary to launch long projectile in order to obtain such pulse duration of few milliseconds, what is a very difficult task, in particular for the realisation of a plane impact of the projectile on the input bar, an absolute necessary condition for the generation of the elastic stress plane wave pulse needed for a correct analysis of the SHPB by means of the one-dimensional elastic plane stress wave propagation theory.

The main modification to the classical split Hopkinson pressure bar consists in the substitution of the projectile, normally used to generate the impact loading pulse, with a statically elastic pre-tensioned bar which is the physical continuation of the input bar as shown in figure 1, where different configurations, for testing materials in tension, compression and shear, are shown.

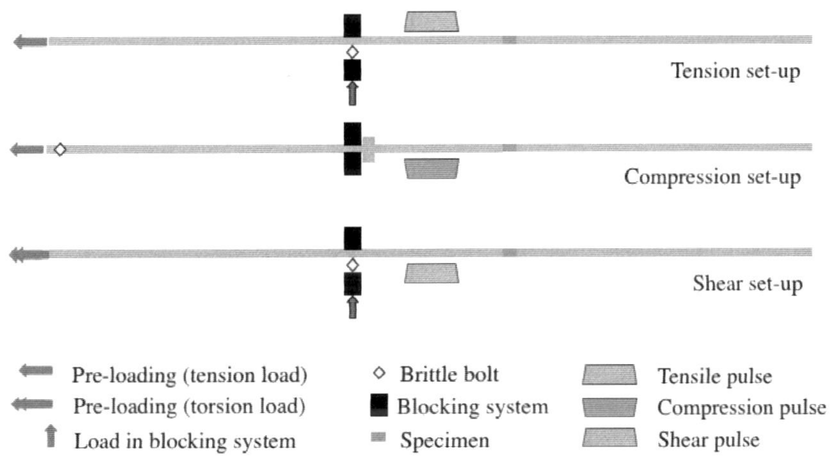

Figure 1. Functioning schemes of the JRC-Modified Hopkinson Bar device.

These devices were used for decades in the analysis of the nuclear materials under harsh conditions as high temperature, high strain rate, irradiation etc. [12].

The way of functioning the JRC-modified Hopkinson bar, consists of the different phases: firstly, elastic energy is stored in the pre-tensioned loading bar by statically tensioning (or twisting) the length of this bar; secondly, by suddenly rupturing a brittle intermediate piece placed in the blocking device a rectangular stress wave pulse is generated and propagates through the input bar, the specimen and the output bar, provoking a state of stress in the specimen till fracture; finally, signals of the input $\varepsilon_I(t)$,

reflected $\varepsilon_R(t)$ and transmitted $\varepsilon_T(t)$ pulses are recorded by the strain-gauge stations glued on the input and output bars which remain in the elastic regime.
Having realised the condition of the specimen deformation in a homogeneous stress state (that means having kept the specimen gauge length sufficiently short in a way that the wave reflections inside the gauge length provide the stress homogenization), and by the application of the elastic one-dimensional stress wave propagation theory, average stress σ, strain ε, and strain-rate $\dot{\varepsilon}$ in the specimen can be determined with the following relationships:

$$\sigma(t)= E\cdot\frac{A}{A_0}\cdot\varepsilon_T(t) \quad (1)$$

$$\varepsilon(t)= -\frac{2\cdot C_0}{L}\int_0^t \varepsilon_R(t)\,dt \quad (2)$$

$$\dot{\varepsilon}(t)= -\frac{2\cdot C_0}{L}\cdot\varepsilon_R(t) \quad (3)$$

where: L = gauge length of the specimen, A_0 = cross-sectional area of the specimen, t = test time, E = elastic modulus of the bar, A = cross-sectional area of the bar and C_0 = the elastic wave speed in the bars.

3. Extreme conditions of high temperature and high strain rate

The mechanical characterization of materials under harsh condition of strain rate and temperature can be obtained adding to the MHB a heating system [13-14]. The functioning scheme is shown in figure 2.

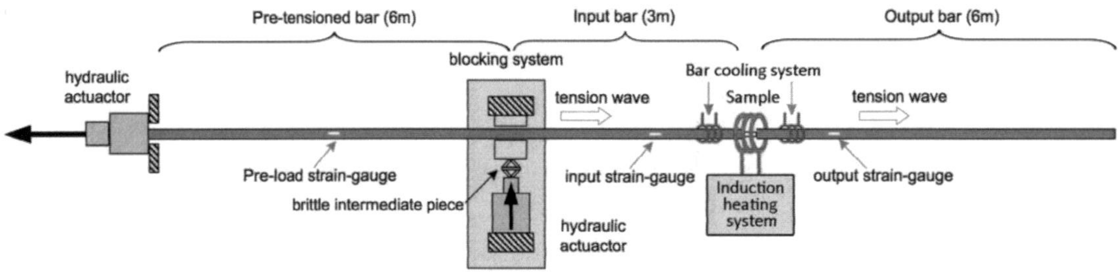

Figure 2. Functioning schemes of the JRC-Modified Hopkinson Bar device with heating systems [13].

3.1. Austenitic stainless steels

The austenitic stainless steels AISI304 (EN 1.4301) was characterized in a wide interval of strain rates [15]. The effects of the high temperatures on its dynamic behaviour was the object of another study [16]. High temperatures were obtained by using an Ambrell compact EASYHEAT induction water-cooled heating systems with maximum power of 2.4kW (see figure 3) [13]. Round samples having 3 mm in diameter and 5 mm of gauge length were used [13-16]. The composition of the austenitic steel AISI304 here examined is shown in Table 1. It has fully austenitic microstructure and contains a large amount of alloying elements as Cr and Ni in order to improve pitting and corrosion resistance.

Table 1. Chemical composition (wt.%) of AISI304 (EN 1.4301)

C	S	P	Si	Mn	Cr	Ni	Mo	Cu	Sn	Al	Co	Nb	N2	Fe
0.025	0.003	0.032	0.370	1.200	18.26	8.10	0.34	0.880	0.010	0.004	0.100	0.020	0.053	Bal.

In figure 3, the setup for the high strain rate tests at elevated temperature is shown. It is possible to observe, the input ① and the output bars ② of the SHTB, the heating system ③, the water-cooled induction coil ④, the sample to be tested ⑥ connected by means of a thermocouple to a thermal controller ⑤ and the cooling system ⑦ for the input and output bars, respectively.

Figure 3. Experimental set-up for high strain rate combined with high temperature.

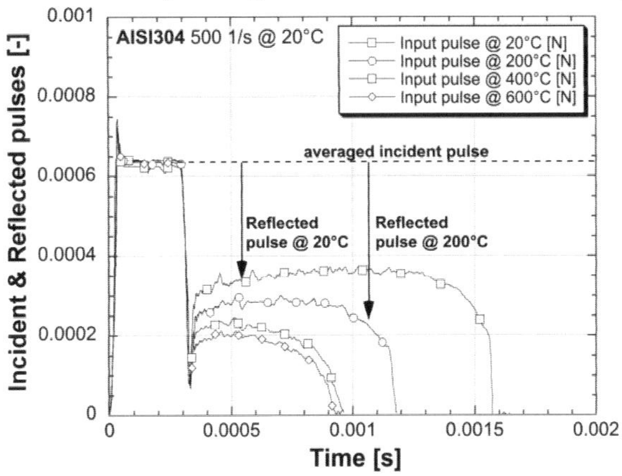

Figure 4. Functioning schemes of the JRC-Modified Hopkinson Bar device.

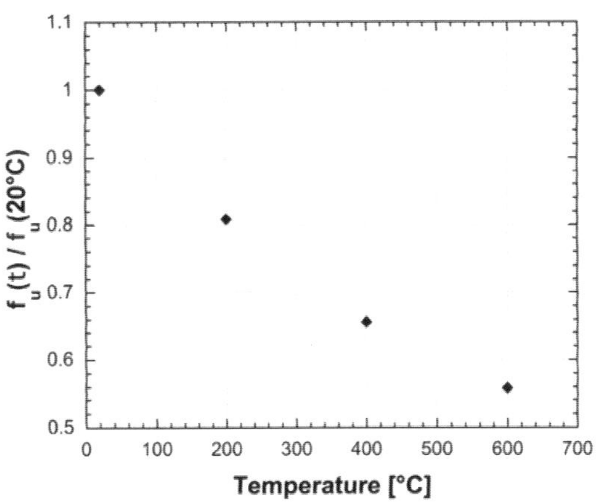

Figure 5. Reduction factor of the ultimate tensile strength at high strain rate in function of the temperature.

In figure 4 the influence on the reflected pulses is depicted highlighting their variation in function of the different temperatures. The influence of the temperature can be observed in figure 5 where the decrement of the ultimate tensile strength in function of the temperature is shown.
Finally, the stress versus strain diagrams are depicted in figure 6 both in terms of engineering and true values.

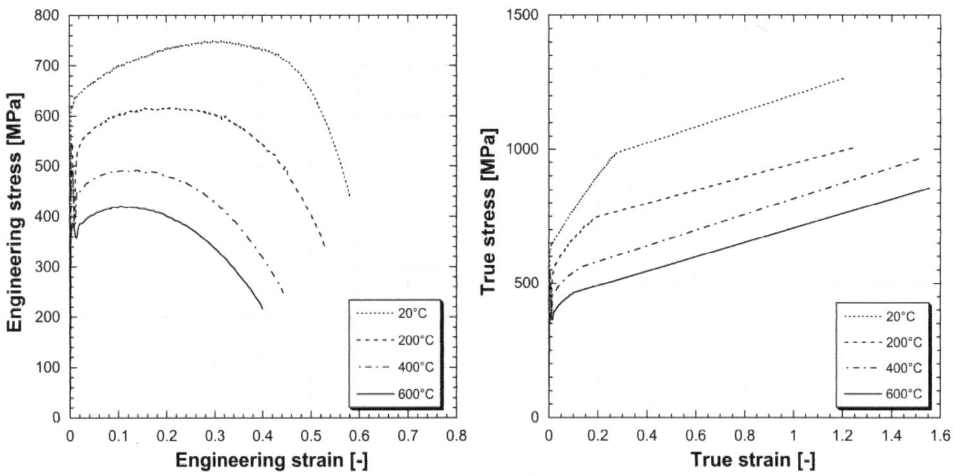

Figure 6. Engineering versus strain (left) and true stress versus true strain (right) diagrams at different temperatures.

3.2. Material constitutive relationships
The material constitutive relationship is a formulation that describe the flow stress as a function of strain, strain-rate and temperature. There are two families of functions. The first is directly obtained by best fitting of experimental values and often expressed as a product of different member each describing only one effect $\sigma(\varepsilon, \dot{\varepsilon}, T) = f(\varepsilon) \cdot g(\dot{\varepsilon}) \cdot h(T)$. The most famous model was proposed by Johnson and Cook [18]. The second one the flow stress is obtained as addition of different members representing respectively strain hardening, strain-rate and thermal softening. The most popular expression was proposed by Zerilli and Armstrong [19] in the form $\sigma(\varepsilon, \dot{\varepsilon}, T) = f(\varepsilon) + g(\dot{\varepsilon}) + h(T)$ and is physical based and are consistent with micromechanics of material behaviour (dislocation density etc). These equations need a certain number of experiments at different strain-rate, different temperature, in order to calibrate the model on the material. Higher is the precision of the results higher is the accurateness of the numerical provision obtained.
As example for this austenitic stainless steel the parameter of the Johnson and Cook relationship were obtained from the experimental results. This is a widely used model, based on three phenomena, the isotropic hardening, the strain-rate hardening and the thermal softening.

$$\sigma_{JC} = \left(A + B \cdot \epsilon_p{}^n\right) \cdot \left(1 + C \cdot \ln {}^{\dot{\epsilon}_p}\!/_{\dot{\epsilon}_0}\right) \cdot \left(1 - \hat{T}^m\right) \tag{4}$$

The parameters to be determined from the experimental results are A, B, and n in order to take into account the strain-hardening, as well as C and m that represent the strain-rate sensitivity and the thermal softening respectively. The previously mentioned material parameters are reported in Table 2.

Table 2. Johnson-Cook parameters for AISI304 (EN 1.4301) steel

A [MPa]	B [MPa]	N [-]	C [-]	m [-]
559	1924	1.040	0.00710	0.87

3.3. High-chromium tempered martensitic reduced activation steel Eurofer97

Eurofer97 has been selected by EU as reference structural material and will be used to fabricate the Test Blanket Modules of the International Thermonuclear Fusion Reactor (ITER). It is a reduced activation steel and it is a tempered martensitic stainless steel of the 7–9wt% Cr class produced by Böhler company. The concentration of the main alloying elements is (in wt.%) 0.1% C, 8.9% Cr, 1%W, 0.2%V and 0.15% Ta (see Table 3).

Table 3. Chemical composition (wt.%) of Eurofer97

C	S	P	Ta	Mn	Cr	V	W	Cu	Ti	Al	Co	N	Si	Fe
0.11	0.001	0.0001	0.12	0.55	8.95	0.2	1.06	0.005	0.001	0.009	0.004	0.022	0.03	Bal.

In order to obtain the reduced-activation behaviour, several alloying elements commonly added to commercial martensitic stainless steels like Ni, Nb and Mo have been either removed (Ni), or replaced (Nb and Mo) by elements with shorter half-lives (W, V, Ta) [20]. This steel is very clean and its inclusions level is extremely low. The mechanical properties are the following: E=217GPa, ν=0.3, ρ=7750 kg/m^3; σ_y=546 MPa; σ_u=668 MPa.
The strain rate effect on the stress versus strain diagram as well as the effect of high and low temperature can be observed in figure 7 and 8.

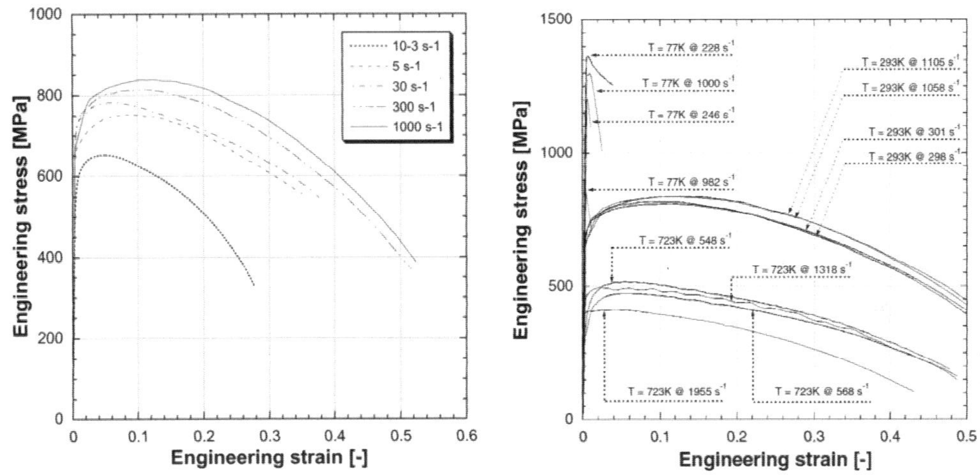

Figure 7. Engineering stress vs strain curves at room temperature and different strain rates (left) engineering stress vs strain curves at room, high and low temperature at different strain rates (right).

The influence of the temperature on the dynamic behaviour of Eurofer97 can be described by the Zerilli-Armstrong relationship for the body centered cubic (bcc) metal. The relationship proposed for bcc metals is:

$$\sigma_{ZA} = c_0 + c_1 \cdot exp(-c_3 \cdot T + c_4 \cdot \ln(\dot{\varepsilon})) + c_5 \varepsilon^n \tag{5}$$

where, c_1 , c_3, c_4, c_5 and n are the five material constants for the bcc model, while c_0 take into account the influence of the dislocation density on the yield stress [21].

Table 4. Zerilli-Armstrong parameters for Eurofer97 steel

c_1 [MPa]	c_3 [°K^{-1}]	c_4 [°K^{-1}]	c_5 [MPa]	n [-]
1075	$2.063 \cdot 10^{-}$	$6.762 \cdot 10^{-}$	673.1	0.5320

5	5

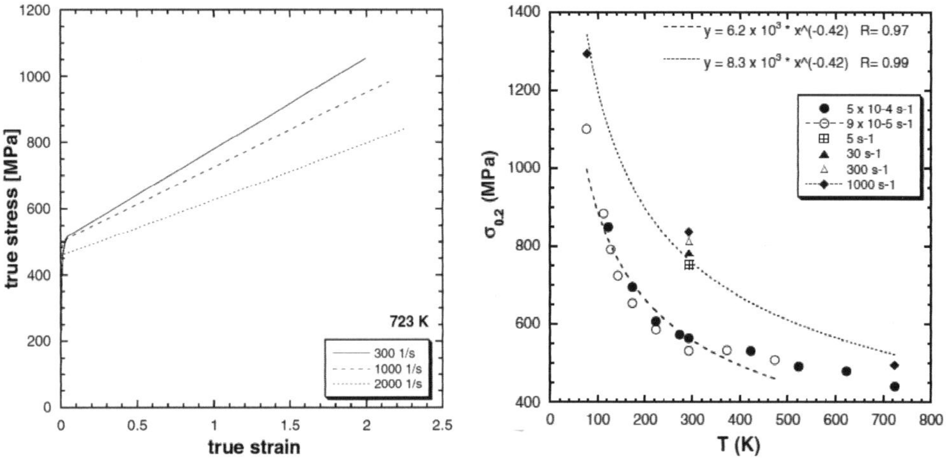

Figure 8. Effect of the strain rate on true stress vs strain curves at room temperature (left) effect of the temperature and strain rate on the yield stress (right) [21].

4. Concluding remarks

In order to analyse the mechanical behaviour of steel alloys at high strain rate and high temperature has been presented the experimental set-up needed for the study of their dynamic behaviour. The experimental results of two steel alloys such as austenitic stainless steel and high-chromium tempered martensitic reduced activation steel Eurofer97 obtained in extreme conditions of high strain rate and high temperature have been presented and discussed.

These results are in the direction of better understanding of these materials under harsh conditions and have allowed to obtain the description of the materials in terms of numerical model [17] and material relationships [15, 16, 20, 21].

This analysis is required whenever a Critical Raw Material has to be replaced by an alloy, especially if it is subjected to extreme temperature and deformation rates. Future projects will be addressed to the study of the effects of the reduction or substitution of Critical Raw Materials in the manufacturing, transport and energy industries in such extreme condition of loading and temperature.

5. References

[1] Albertini C. and Montagnani M 1974 Testing techniques based on the Split Hopkinson bar *Inst. of Physics, Conf. Ser. No 21, London,* 22–32.

[2] Albertini C and Montagnani M1976 Waves Propagation Effects on Dynamic Loading *J Nucl Engg Des* **37** 115–24.

[3] Cadoni E, Albertini C and Solomos G 2006 Analysis of the concrete behaviour in tension at high strain-rate by a modified Hopkinson bar in support of impact resistant structural design, *J de Phys IV* **134** 647–52.

[4] Fenu L, Forni D and Cadoni E 2016 Dynamic behaviour of cement mortars reinforced with glass and basalt fibres, *Comp Part B: Engg* **92**(5) 142–50.

[5] Cadoni E, Labibes K, Berra M, Giangrasso M and Albertini C 2001 Influence of Aggregate Size on Strain-Rate Tensile Behavior of Concrete, *ACI Mat J* **98**, 220–3.

[6] Asprone D, Cadoni E, Iucolano F and Prota A 2014 Analysis of the strain-rate behavior of a basalt fiber reinforced natural hydraulic mortar *Cem Conc Comp* **53**(10) 52–8.

[7] Cadoni E, Solomos G and Albertini C 2009 Mechanical characterisation of concrete in tension

and compression at high strain rate using a modified Hopkinson bar, *Mag Conc Res* **61** 221–30.
[8] Asprone D, Cadoni E, Prota A and Manfredi G 2009 Strain-Rate Sensitivity of a Pultruded E-Glass/Polyester Composite, *ASCE–J Comp Const* **13**(6) 558–64.
[9] Coppola L, Cadoni E, Forni D and Buoso A 2011 Mechanical Characterization of Cement Composites Reinforced with Fiberglass, Carbon Nanotubes or Glass Reinforced Plastic (GRP) at High Strain Rates, *Appl Mech Mat* **82** 190–5.
[10] Cadoni E and Forni D 2016 Experimental analysis of the UHPFRCs behavior under tension at high stress rate, *Eur Phys J Spec Top* **225** 253–64.
[11] Cadoni E, Singh N, Forni D, Singha M and Gupta N 2016 Strain rate effects on the mechanical behavior of two Dual Phase steels in tension', *Eur Phys J Spec Top* **225** 409–20.
[12] Albertini C, Cadoni E and Solomos G 2014 Advances in the Hopkinson bar testing of irradiated/non-irradiated nuclear materials and large specimens, *Phil Trans Roy Soc A* **372** 20130197.
[13] Forni D, Chiaia B and Cadoni E 2016 High strain rate response of S355 at high temperatures, *Mat Des* **94** 467–78.
[14] Forni D, Chiaia B and Cadoni E 2016 Strain rate behaviour in tension of S355 steel: Base for progressive collapse analysis *Engg Struct* **119** 164–73.
[15] Cadoni E, Fenu L, and Forni D 2012 Strain rate behaviour in tension of austenitic stainless steel used for reinforcing bars, *Constr Build Mat* **35** 399–407.
[16] Cadoni E, Dotta M and Forni D 2016 Tensile behaviour of reinforcing steels at high strain rate and high temperature *Key Engg Mat* **711** 791–8
[17] Riganti G and Cadoni E 2014 Numerical simulation of the high strain-rate behavior of quenched and self-tempered reinforcing steel in tension *Mat Des* **57** 156–67.
[18] Johnson GJ and Cook WH 1985 Fracture characteristics of three metals subjected to various strains, strain rates, temperatures and pressures. *Engg Fract Mech* **49** 21–31.
[19] Zerilli FJ and Armstrong RW 1987 Dislocation-mechanics-based constitutive relations for material dynamics calculations *J Appl Phys* **61** (5) 1816
[20] Cadoni E, Dotta M, Forni D and Spaetig P 2011 Strain-rate behavior in tension of the tempered martensitic reduced activation steel Eurofer97, *J Nucl Mat* **414**(3) 360–6.
[21] Cadoni E, Dotta, M, Forni D and Spaetig P 2015 Influence of the temperature on the tension behaviour of EUROFER97 alloy at high strain rate *EPJ WoC* **94** 01022.

High temperature properties of non-critical Fe-Al alloys doped by non critical or "slightly–critical" elements

Věra Vodičková[1*], Pavel Hanus[1], Tomáš Vlasák[2], Martin Švec[1]

[1] Technical University of Liberec, Studentska 1402/2, Liberec 1
[2] The Laboratory of Material Properties at High Temperatures, Podnikatelska 565, Prague

*corresponding author: vera.vodickova@tul.cz

Abstract. Iron aluminides were developed as an alternative to stainless steels after World War II. The main intended impact was to save strategic elements (chromium or nickel). The result of these investigations was development of registered alloys as Pyroferal (Czechoslovak Republic), Thugal (Soviet Union) or Thermagal (France). The investigation of these type alloys continued in the nineties thanks to technological progress. In this time iron aluminides seems to be promising material with very good corrosive and environment resistivity. The mechanical properties of binary iron aluminides (Fe-Al) are average at higher temperatures but strengthening effect of alloying elements is significant. The aim of the article is to show influence of non-critical additives (such as C, Ti, Zr) and also "slightly critical" elements as e.g. Ce, Nb on high temperature creep properties of alloys.

1. Introduction

Iron materials alloyed by aluminium as heat resistant materials were known even before 1900 [1]. In the thirties of the 19th century the investigations were focused on the effect of Al adding to cast iron [2, 3].

The long-term interest for aluminium modified iron materials started in Europe scale after World War II. The initial focus on the effect of adding Al to white and grey cast iron was replaced by the idea to substitute chromium and nickel in heat resistant cast iron in later studies. These efforts resulted in large-scale production of ternary Fe-Al-C materials. The result of these investigations was development of new materials such as Tchugal in former Soviet Union, Thermagal in France or Pyroferal in former Czechoslovakia.

Pyroferal alloy was the result of very detailed research of such type of materials initiated in Czechoslovakia in the 1950s of the last century. The quasi-binary phase diagrams [4,5] for Fe-xAl-yC were constructed based on the investigation of the phases in the Fe-Al-C system. Pyroferal alloy [6] exhibited good mechanical properties at higher temperatures and especially very good corrosion resistance (the layer of aluminium oxide that was formed protected the material comparatively better as the substituted materials highly alloyed by Ni and Cr). Also resistance against aggressive environment (against the atmosphere of the natural gas cracking generator or against molten glass) or carburization was tested with very successful results.

Recently the „like Pyroferal" materials were tested again to find out the reasons for the excellent H.T. corrosion resistance and mechanical properties [7, 8]. Namely the dependence of the creep properties and structure at 600°C on the content of carbon and silicon was tested [9].
On a global scale the intensive research focused on aluminides was started in the late nineties. The alloys with FeAl and Fe_3Al matrix were also investigated in Czech Republic, especially at the Charles University in Prague, in cooperation with research teams in Liberec, Brno and Ostrava.

2. Improvement of high-temperature mechanical properties of Fe_3Al-type Iron Aluminides

In general, iron aluminides exhibit excellent corrosion and environment resistance, they have low density and – important attribute for industrial use – they are affordable. Fe_3Al-type iron aluminides with $D0_3$ lattice have good strength at moderate temperatures due to high degree of crystallographic order.

Current research is focused on improving high-temperature (H.T.) tensile and creep properties of Fe_3Al-type iron aluminides. The task to enhance the H.T. mechanical properties of Fe_3Al based alloys can be arranged by using of different methods [10]:
- strengthening by solid solution hardening which is useful for addition of small amount of ternary elements such as Cr, V, Mo and Ti,
- strengthening by incoherent precipitates which may be either formed due to ternary additions of Zr, Ta, Nb or Ti in higher concentrations (diverse phases originate according to high-temperature isothermal section of phase diagrams) or carbide and boride precipitates, which need additions of C or B (also as quaternary to the above mentioned elements),
- strengthening by coherent precipitates as one of the most effective methods (B2 in A2 and vice versa for Fe-Al-Ni-Cr alloys or B2+$L2_1$ in Fe-Al-Ta system),
- strengthening by increased crystallographic order which stabilizes the $D0_3$ structure to higher temperatures (e.g. Fe-Al-Ti).

Another alternative to increase H.T. mechanical properties is represented by directionally solidified eutectic alloys, Fe-Al-Nb for example, that showed a higher creep resistance only at 650°C – 750°C compared with the as-cast alloys of about the same composition. The creep resistance at higher temperature remained at the same level [11].
An especially effective method is based on the addition of elements with a low solubility in the Fe Al-matrix. From this point of view, the alloying by zirconium may be used. The beneficial effect of zirconium addition on creep resistance was first written by McKamey and Maziasz [12, 13]. The incoherent phases are typically formed in iron aluminides without carbon addition, i.e. pure ternary Fe–Al–Zr alloy. Even small amounts of Zr leads to the precipitation of Laves-phase $(Fe,Al)_2$ Zr or τ_1-phase $(Fe,Al)_{12}$ Zr particles that may limit of dislocation motion during deformation.

Zirconium was tested as an additive to Fe_3Al-type iron aluminide to enhance the H.T. tensile by Kratochvíl et al. [14,15] The formation of ZrC particles, which con- tributed to improvement of tensile creep properties, was described. Kratochvíl et al. [16] compared influence of Zr combined with TiB_2 on creep resistance of Fe_3Al-type alloy with chromium content. Significant influence of Cr (strengthening by solid solution hardening) was described by McKamey in [17]
In Fig. 1, the comparison of Minimum Creep Rates (MCR) at 600°C is shown for alloys doped by Zr combined with TiB_2 (pink areas) investigated in [16], doped by Zr (FA6, FA11) [14], by TiB_2 (FA3) and by Ce (FA1) [18]. It was demonstrated, that the combination of additives Zr + TiB_2 in the tested alloy does not improve the creep resistance. The material slightly alloyed with Zr with fine dispersion of ZrC particles is obviously the best of the compared iron aluminides [16]. Therefore, Zr was confirmed as suitable element for enhancement of H.T. mechanical properties of Fe_3Al-based iron aluminides.

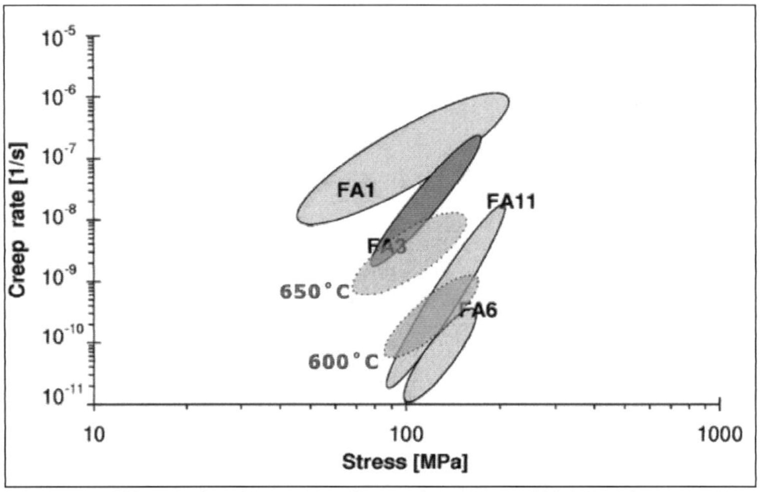

Figure 1. The comparison of MCR at 600°C [16]

If the low cost of the raw materials is required, the presence of carbon in raw Fe is unavoidable. Carbon can be also used as one of the cheapest available alloying elements, with its low solubility in iron and a negligible solubility in aluminium. Moreover, carbon content appears to be beneficial in suppressing the hydrogen embrittlement at grain boundaries.The research focused on carbon addition [19, 20] confirmed that the influence of carbides morphology and distribution on the properties of alloys with 24 - 28 at. % Al can be significant. Baligidad et al. [19] attributed the improvement in the mechanical properties at elevated temperature to the presence of the Fe_3AlC phase (carbide κ) as well as the interstitial carbon in the matrix. On the other hand, according to Schneider et al. [20], the carbide κ does not enhance of the H.T. strength efficiently due to fast coarsening.
While the influence of zirconium or carbon on alloy strengthening has been described in great detail, less attention has been devoted to the combined additions of zirconium and carbon. This summarizing article purpose is to present a part of research focused on the common effect of zirconium and carbon with the aim to find an optimal concentration ratio of the additive elements to increase H.T. mechanical properties.

3. The effect of combined addition of zirconium and carbon on creep resistance of Fe_3Al-based alloys - results and discussion

The effect of phase composition on H.T. mechanical properties of Fe_3Al-based iron aluminides **with chromium addition** (3.7 at. %) was investigated through H.T. properties testing and demonstrated via microstructure analysis [21]. The alloys were prepared by vacuum induction melting and casting, the casts were rolled at 1200°C.

Phase composition of low-alloying iron aluminides (Zr content 0.2-0.3 at. %, C content 0.2-0.6 at. %) was investigated by means of scanning electron microscopy and energy dispersive X-ray analysis and subsequently verified through electron diffraction by TEM. Creep tests with stepwise loading in temperature range 600 – 800°C were used. Tests were carried out in a protective atmosphere of dry purified argon. Creep resistance was compared through Minimum Creep Rate values [21]. This paper studied the effect of the c(Zr)/c(C) ratio on the microstructure and H.T. mechanical properties of Fe_3Al-type aluminides. Two cases are distinguished: materials with Zr excess and those with C excess. Based on comparison of phase composition of alloys alloyed with Zr and C simultaneously, the relative ratio of zirconium-carbon concentrations can play a significant role. With respect to the ratio c(Zr)/c(C), there may be two cases of phase formation:
- with the carbon predominance, the zirconium affinity toward the carbon prefers the formation of carbides ZrC, and it is the majority phase.

- with the Zr predominance, ZrC is again preferred (due to the high affinity of Zr to C), but after consumption of carbon, there will be enough Zr to form of the Laves phase.

An exact knowledge of nature of the secondary phase particles can contribute to the improvement of creep resistance at high temperatures. This research confirmed that the concentration ratio of two additives (zirconium and carbon) influenced the phase structure of the investigated undeformed alloys in agreement with the relationships in the Fe–Al–Zr ternary phase diagram.The secondary phase particles - Laves phase λ_1 $(Fe,Al)_2$ Zr and ZrC - were distributed primarily along grain boundaries, with some distribution in the grains as well – Fig. 2. Their strengthening role is limited to blocking the grain boundaries' motion because the effect of these particles inside the grains (owing to large distance between the particles) is insignificant. During the H.T. deformation two different metastable phases $(Fe_{1-x}Al_x)_3Zr$ and Fe_2Zr originate and interact with dislocations. The changing density and size of these particles with temperature explains the described change in the creep behavior [21].

The combined effect of carbon and zirconium on creep resistance can be evaluated from the point of view of the concentration ratio c(Zr)/c(C). The stress for producing a creep rate of 10^{-7} s^{-1} is presented as a function of this ratio - Fig. 3. An increase of this stress with c(Zr)/c(C) > 1 is in agreement with contribution of the Laves phase to creep resistance [21].

In follow-up research [22, 23, 24] on the influence of ratio was carried for **Cr-free** iron aluminides. The examined alloys were prepared by vacuum melting and casting; Zr was added in metallic form.

Due to higher amount of zirconium in the investigated alloys, Laves phase was present as eutectic mixture with matrix. It has been demonstrated that the relative ratio of carbon-zirconium concentrations affects the amount of Zr remaining for Laves phase formation, which subsequently affects volume fraction of Laves phase. Finally, resulting H.T. mechanical properties are influenced by volume fraction of the eutectic mixture of Laves phase and matrix. The influence of the volume fraction of the eutectic (Laves phase and matrix mixture) on the strengthening was confirmed through H.T. yield stress in compression for materials with minimum amount of carbon (Fig. 4) [22] as well as materials with higher carbon (Fig. 5) [23].

The effect of zirconium/carbon ratio on creep resistance of chromium-free iron aluminides was also investigated [24]. The alloys with higher amount of zirconium and different carbon contents (0.06 to 3.46 at. %) were tested. The examined alloys were prepared by vacuum melting and casting. The samples were subjected to stepwise loading, where the load was changed after the steady-state creep

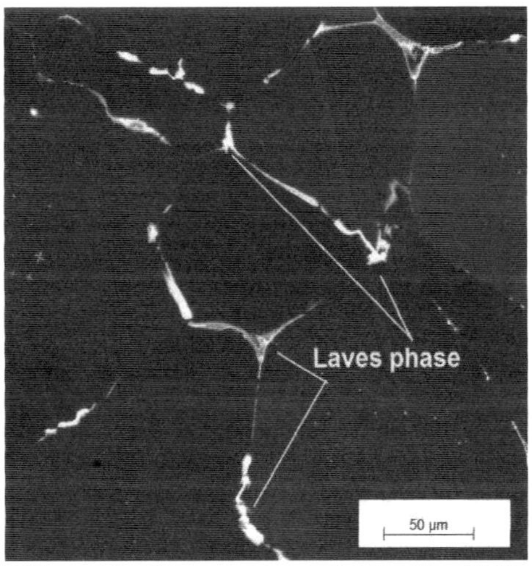

Figure 2. The structure of 1A alloy [21]

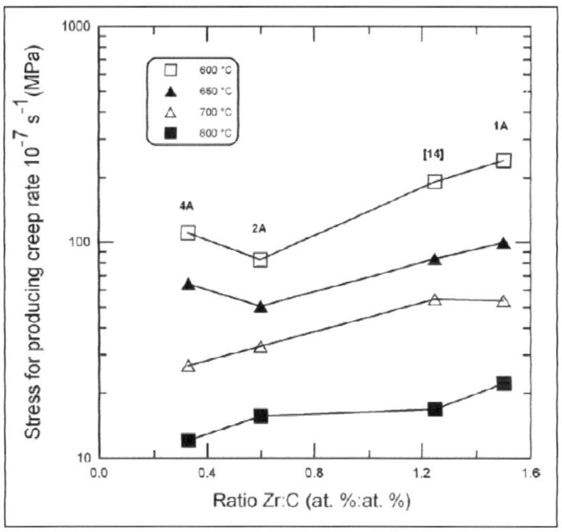

Figure 3. The stress necessary to produce a creep rate of 10^{-7} s^{-1} as a function of c(Zr)/c(C) [21]

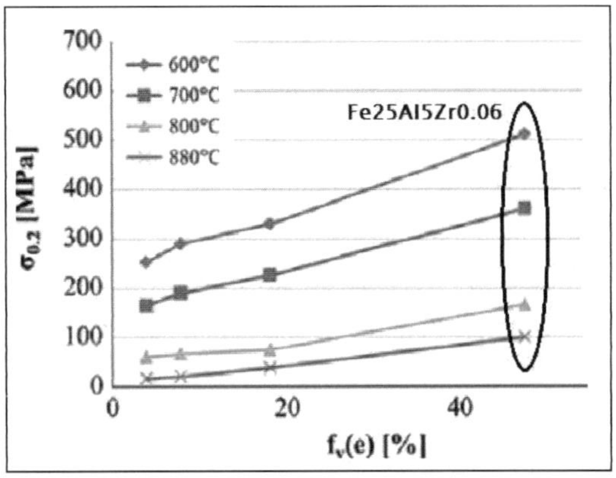

Figure 4. Dependence of $\sigma_{0.2}$ of the as-cast Fe-25Al alloys on the volume fraction of the eutectic $f_v(e)$ at various temperatures [22]

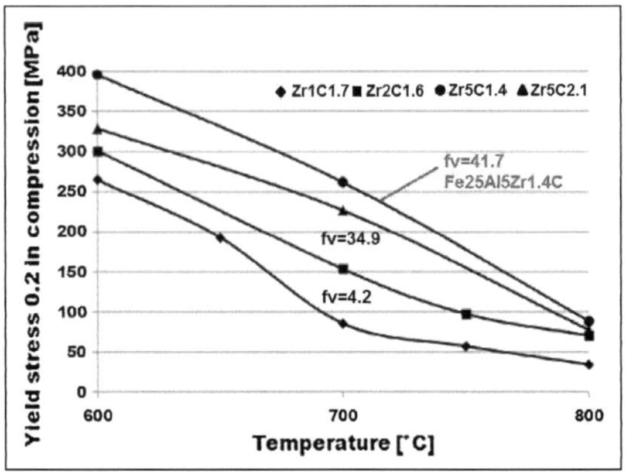

Figure 5. The dependence of the yield stress $\sigma_{0.2}$ on temperatures for as-cast alloys [23].

rate was established for a given load. Creep resistance was evaluated using dependence of the creep rate on the applied stress σ at the various temperatures. Stress exponent n and activation energy of creep were compared. Results were also discussed in terms of the concentration ratio of c(Zr)/c(C). In Fig. 6, the stress for producing a creep rate of 10^{-7} s^{-1} is presented as a function of this ratio again. It is obvious that an increase of the creep resistance with increasing ratio Zr/C is in agreement with the behavior observed previously in alloys with lower concentrations of Zr and with Cr addition [21].

The effect of concentration ratio corresponds with the relationships in the Fe–Al–Zr ternary phase diagram (and its potential extension by adding carbon). It was confirmed that the presence of C reduces the amount of Zr that is available for the formation of the Laves phase and/or other Zr-phases that is why the creep resistance at 973 K (700 °C) is reduced [24].

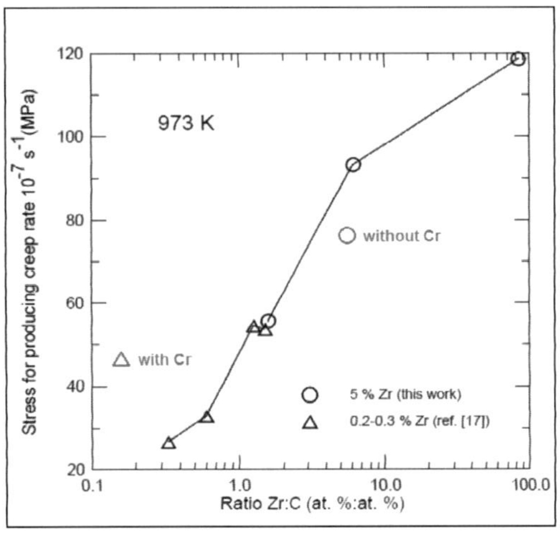

Figure 6. The stress necessary to produce a creep rate of 10^{-7} s^{-1} as a function of c(Zr)/c(C), [17] in Fig = [21] in this paper

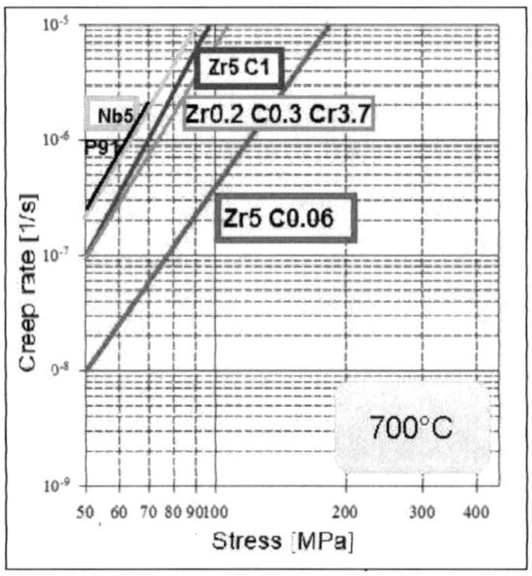

Figure 7. The comparison of creep rates at 700°C of alloys mentioned in this work

The comparison of creep rates of alloys mentioned in this work is given in Fig 7. Alloy Zr0.2C0.3Cr3.7 represents the material with Cr addition tested in [21]. Alloys Zr5C0.06 and Zr5C1 are chromium-free materials investigated in [24]. The creep rate of alloy with 5 at. % Nb [25] is shown, because there is similarity between Fe-Al-Zr and Fe-Al-Nb systems especially with regard to region of Laves phase. In the Fe-Al-Nb system, the strengthening effect of the solid solution hardening must also be taken into account [26]. The values of P1 steel creep rate [27] are given as a representative of alloys with good creep properties. Based on this comparison it can be assumed that strengthening of Fe_3Al-type iron aluminides by chromium (critical element) - i.e. strengthening by solid solution hardening - can be fully replaced by the Laves phase strengthening due to alloying with zirconium (non-critical element), taking into account the "a priori" presence of carbon in the alloy.

4. Conclusion

With simultaneous alloying with zirconium and carbon, the ratio of their concentrations must be taken into account. As well as with zirconium alloying, if common quality iron is used for production of alloy.

The present of carbon always reduces the amount of zirconium available to produce the reinforcing Laves phase. By increasing the concentration ratio c(Zr)/c(C), creep resistance increases.

This ratio also affects the amount of Zr that may contribute to the formation of a fine dispersion of the particles of other phases during creep deformation. These fine particles can act the function of obstacles to move the dislocations.

References

[1] Borsig A. Stahl Eisen 1894:6.
[2] Everest AB. Foundry Trade J 1927;37:208.
[3] Thyssen H. J Iron Steel Ind 1934;130:153
[4] Pluhař J and Vyklický M Properties of the heat resistant alloy CSN 42 2484 (Pyroferal) and its use. Technical report of the heavy industry, No. 189. Prague; 1959 [in Czech].
[5] Vyklický M and Tůma H Hutnicke listy 1959;14:118 [in Czech]
[6] Czechoslovak Standard No. 422484-Alloy Pyroferal.
[7] Vodičková V, Kratochvíl P, Dobeš F, Kovove materiály - Metallic Materials 45 (2007) 153
[8] Kratochvíl P, Dobeš F, Vodičková V, Intermetallics 17 (2009) 39
[9] Dobeš F, Kratochvíl P, Vodičková V, Intermetallics 19 (2011) 1526
[10] Palm M, Intermetallics 13 (2005) p. 1286
[11] Milenkovic S, Palm M, Intermetallics 16 (2008) p. 1212
[12] McKamey CG, Maziasz PJ, Jones JW J. Mater. Res. 7 (1992) 2089–2106. 1172
[13] McKamey CG, Maziasz PJ, Goodwin GM, Zacharia T, Mater. Sci. Eng. A174 (1994) 59
[14] Kratochvíl P, Málek P, Cieslar M, Hanus P, Hakl J, Vlasák T, Intermetallics 15 (2007) 333
[15] Dobeš F, Kratochvíl P, Milička K, Strength Mater. 40 (2008) 106
[16] Kratochvíl P, Vodičková V, Hanus P, et al., Intermetallics (2010) 1365
[17] McKamey CG, Horton JA, Liu CT, Journal of Materials Research 1989 (4), Issue 5, p. 1156
[18] Kratochvíl P, Málek P, Pešička J, et al., Metallic Mat., (2006) 44, p. 185
[19] Baligidad RG, Prakash U, Radha Krishna A. Mater Sci Eng 1997;A230:188.
[20] Schneider A, Falat L, Sauthoff G, Frommeyer G. Intermetallics 2005;13:1322
[21] Kratochvíl P, Dobeš F, Pešička J et al, Mater Sci Eng A 548 (2012) 175
[22] Kejzlar P, Kratochvíl P, Král R, Vodičková V, Metallurgical and Materials Transactions 45A (2014) 335
[23] Kratochvíl P, Vodičková V, Král R, Švec M, Metallurgical and Materials Transactions 47A (2016) 1128
[24] Dobeš F, Vodičková V, Veselý J, Kratochvíl P, Metallurgical and Materials Transactions 47A (2016) 6070
[25] Švec M, Dissertation, Technical University of Liberec, Liberec, 2016
[26] Prymak O, Stein F, Intermetallics 18 (2010). 1322
[27] Vlasák T, Hakl J, Novák P, Sochor J, Čech J, Metal 2013 Conference proceedings, ISBN 978-80-87294-39-0, Brno, 2013

IOP Conf. Series: Materials Science and Engineering **329** (2018) 012008 doi:10.1088/1757-899X/329/1/012008

Educating T-Shaped professionals to meet substitution challenges and developing business models for substitution and recycling

Ana Arroyo[1], Javier Mendibil Eguiluz[1], Laura Sánchez Cupido[1]

[1]Tecnalia Research & Innovation, Spain

ana.arroyo@tecnalia.com, javier.mendibil@tecnalia.com, laura.sanchez@tecnalia.com

Abstract. One strategy to overcome the challenges related to critical raw materials (CRMs) is their substitution and recycling. However, the bright scientific idea, proof of concept or laboratory demonstration need to cross the valley of death in order to become stated as 'a substitute' instead of 'a potential substitute'. Most PhD students and Post Docs specialize within a given thematic area; for example on specific materials or on substitution in a certain application. This specialization could limit the ability to generate innovations and profitable business models if there are not enough tools and skills to transform new knowledge and research results into an appealing value proposition towards customers and to a business opportunity for the current markets. The project proposes a framework for developing substitution and recycling related cross-sectorial skills and tools. These are applied for training business-related competences e.g. teamwork, management, communication, value proposition and business models design, especially within RTOs and industries. The proposed learning itinerary can radically improve the path from scientific proof of concept into innovation and lean start up or industrial market launch. The developed framework is tested by a pilot group having several topics within the areas of substitution and recycling of critical raw materials.

1. Introduction

While the concept of T-shaped profile is not so new, it is now more important than ever. The world is becoming more and more complex and T-shaped skills are a razor that can cut through the complexity. Over the past decade, research has emphasized the need for today's young professionals to possess deep disciplinary knowledge along with a keen ability to communicate across social, cultural and economic boundaries. These "T-shaped professionals" are in high demand for their ability to innovate, build relationships, advance research and strengthen their organizations.

Currently many college and university graduates have been trained to be productive in one field, but employers are placing increasing importance on skills that reach beyond a single discipline or focus. Upon graduation, students should be able to handle information from multiple sources, advance professional relationships across different organizations, contribute innovatively to organizational practices, and communicate with understanding across social, cultural, economic and scientific disciplines. Tomorrow's workers will build their careers in a globally interconnected and constantly changing world with smarter technologies in an effort to effect positive global change [1].

A T-shaped skillset must constantly become broader (general knowledge) and narrower (specialized knowledge) as the world becomes more complex, nevertheless having T-shaped skills is one of the biggest competitive advantage an employee can have in the creative society when contributing value for the markets or when achieving goals [2].

In this paper a methodology called "T-DoRe itinerary to develop T-shaped profiles" is presented together with its application to a substitution and recycling case: Neodymium recovery from spent magnets and the corresponding business case.

The definition and implementation of this methodology has been developed under the project T-shaped Doctors in Residence – Doctoral training in RTOs and Industries (T-DoRe) with the support of the European Institute of Technology of Raw Materials, and by a consortium of VTT and Tecnalia Research and Technological Centres, Lappeeranta University and Mondragon University, and the company Outotec.

2. Methodology: T-DoRe itinerary

The situation today is that PhD students and Post Doctorials specialize in certain thematic areas, acquiring a deep technical knowledge. However, they find difficulties when addressing customers' needs, business perspective and lack holistic knowhow that leads to breakthrough innovations.

T-DoRe proposes a general framework based on learning-by-doing methodology to develop T-shaped profiles. This framework describes a global learning itinerary and the main drivers are:

- **Self-development**: It has to do with the capacity to take responsibilities, to take initiative and be proactive, to guide others, to work in teams, to lead and plan a personal and professional project, to learn constantly and to create networks. It is a reflection space for the participants to think about the present and the future needs in order to become T-shaped and think about the type of professional career they want to develop.
- **Ideation for industry and society**. This is about transforming knowledge into innovations. It is about training the capacity to see opportunities and take ideas into action, to value creativity for prototyping solutions (products or services) that can be transferred to the industry, and to experiment in real context with orientation to get results (Value). Bringing research results to the market has to be in the centre of the learning itinerary, and while the participant develops and progress with the project, the research itself is the main tool to create the learning environment for co-operation, thinking out of box, problem solving and innovation.

2.1. Educational approach

The learning methodology proposed in T-DoRe has the following elements:

- **Learning by doing and experimenting.** It means learning by working in real projects, learning from actions done and from direct and frequent contact with industry customers, users, providers and other stakeholders. Learners must be proactive in setting goals, choosing paths and searching for the right tools and skills to develop in each case. It is a pull model (active behaviour of participants) versus push model (passive behaviour as in regular lectures).
- **Learning communities - learning with others**. Learning in teams or in networks/communities allows a faster learning process and the results of the projects may have greater impact. The right environment and context to provide opportunities to learn from regular lectures or traditional learning tools is very important. Participants learn from sharing their own experiences and from others' experiences. The action and reflection model applied continuously provides a change in the learner mind-set, who starts to see any situation related to the project as a real learning opportunity.

While working in a project or thesis, participants acquire knowledge and skills that generate specialists in the raw materials processes and technologies, including the environmental and sustainability aspects. Participant's project is the vehicle to develop transversal/cross skills such as leadership and teamwork, communication, customer development, market validation, prototyping solutions to contrast with industry and see the business perspective to create innovations. And the only way to learn is being led by customers and by end-users. This means the construction of a user-led open innovation ecosystem and especially seeing end-users as an integral part of the innovation process [3].

The learning is based on a participant's project. It can be a research project, an specific project for a company, a PhD thesis itself, or even specific actions needed within regular project activities. The topic of the project is a technical subject related to EIT Raw Materials themes of interest, which are raw materials resource assessment, mining, new mineral and metallurgical processes, recycling methods, substitution of critical and toxic materials and substitutions for optimized performance, and design of products and services for the circular economy [4].

2.2. Structure of the T-DoRe itinerary

The concept of the itinerary is like a journey plan with special places to visit. Within the project a general itinerary is suggested. Depending on the need of the RTO/industry and the background or experiences of the future PhD, this general itinerary may be customized to the participant. It is conceived as learning-by-doing path. Also projects and thesis have different levels of maturity. They could be in their early stage or nearly to be finished. This means that there is one general itinerary but different ways of living it.

While participants are working in the agreed project within the RTO/industry, they have to create new ideas that will lead to innovations. In addition, participants are expected to create the associated value proposition and business model. The lean start up concept and tools together with value proposition and business model approach will be the basis for an iterative project development process.

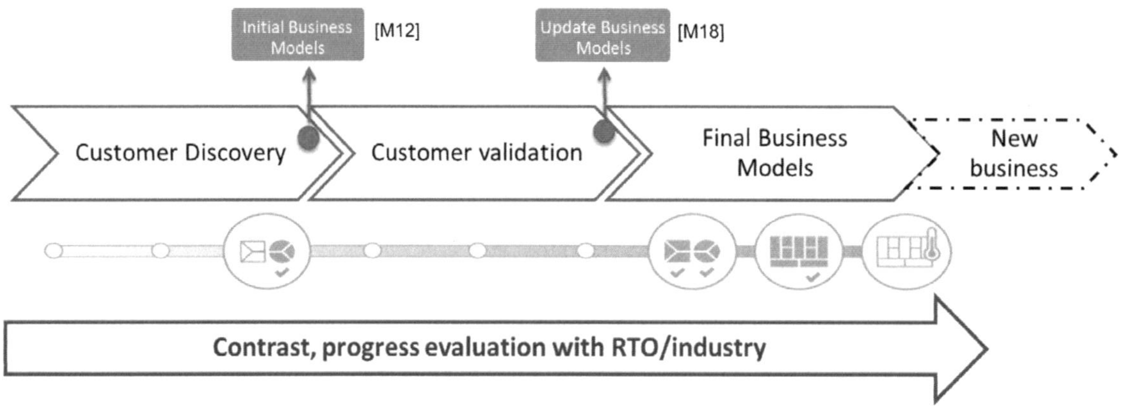

Figure 1. Process to create a business model for the innovations

Figure 2 describes the project development process which evolves and progresses along with the relation with the market. Steve Blank, a Silicon Valley serial-entrepreneur and academician, has described this process in his books. Blank is recognized for developing the Customer Development method that launched the Lean Startup movement. He says that customer discovery is about going out of the building to learn about your customers' jobs, pains, and gains. This has to do with understanding needs and problems of potential customers and investigate what you can offer them to kill pains and create gains. Customer validation is about testing if customers value how products and services intend to alleviate pains and create gains [5].

Customer discovery and customer validation is iterative in a loop; it is a searching activity until "customer creation" where customers are in the sales channels. That is why during all the process there will be contrast and progress evaluation with industry and RTO [5].

Within the scope of T-DoRe this customer creation is represented by a final business model. T-DoRe itinerary produces a business model at the end of the process. Customer visits for feedback, development and prototypes, and customer/user validation is part of the itinerary

During visits to RTOs and industries, participants will have the opportunity to develop or to strengthen the skills defined in their personal learning itinerary. One topic T-DoRe will stress is the frequent contact with Industry. It can be in the form of frequent visit, communication, or in shorts internships.

Based on Blank's customer development approach, a general description of the process is: It starts with the definition of a value proposition hypothesis that should be tested with stakeholders several times; learning from this testing will help to update the business model. Participants during the project have to work on the business models, workshops and coaching sessions are the moment to share the progress with other participants.

The itinerary starts with the presentation of the objectives, the activities to perform, the workshops to attend, how the process will be measured, and the communication dynamics with their mentors in RTO or industry, and with the coaches.

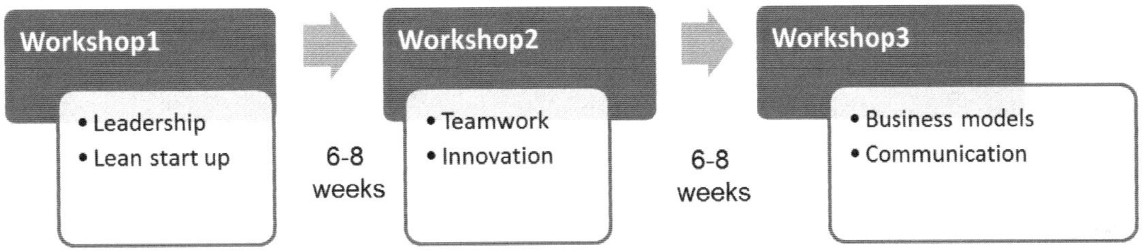

Figure 2. Workshops in the itinerary

Participants will meet at several workshops with the aim of:

- Learning about a specific topic and share progress in their practice/projects.
- Sharing results from their projects and learning from others' experiences.

As shown in Figure 3, workshops focus on specific general themes relevant to all projects, as innovation, marketing and so on. The number of workshops may vary but they normally would occur every 6-8 weeks, giving some time to participants to make progress in the project (customer visit, testing a prototype, etc.).

2.3. Coaching sessions and follow up

At the beginning of the itinerary, participants elaborate a learning contract where their learning plan is stated which represents the seal a commitment to fulfil it. The activities during their project will be guided by this plan.

T-DoRe itinerary proposes coaching sessions between the workshops. The aim of these sessions is to follow up actions, set new objectives if necessary and share practices among participants. It is a way to be connected to the community.

Participants during the project have to work on the business models and workshops and coaching sessions are the moments to share the progress. A monthly 4 hours coaching session, is maintained and it can be done presently or it can be virtual, using Skype or other tool.

3. T-DoRe itinerary in practice

Eleven participants (PhD candidates) have done the itinerary to acquire a T-shaped profile. One important activity of the T-DoRe itinerary has been the elaboration of the Business Case which is a container of the results of actions taken and reflects the progress in the way to bring the research results to the market.

Participants were invited to develop their research project business case. In this sense the 11 participants (PhD candidates) first created a value proposition canvas, covering the main aspects of the value proposition resulted from their research for an specific customer segment.

The value proposition canvas is a simple way to understand the customers' needs, and design products and services they want. This tool was created to help business builders to be very clear about whom a business is for and how it delivers value to them [6].

So for the participants this tool forced them to think like the customer would do and then figure out why their product or service is valuable. Once the customer is selected, they start with the Wants, Needs, Fears and Substitutes in order to understand the problems the customers are willing to solve, and the gains or values the customers would like to achieve. In parallel, participants think about their project main features, benefits that produce and customer experience and satisfaction. This exercise was done supported by the analysis and visits done to potential customers.

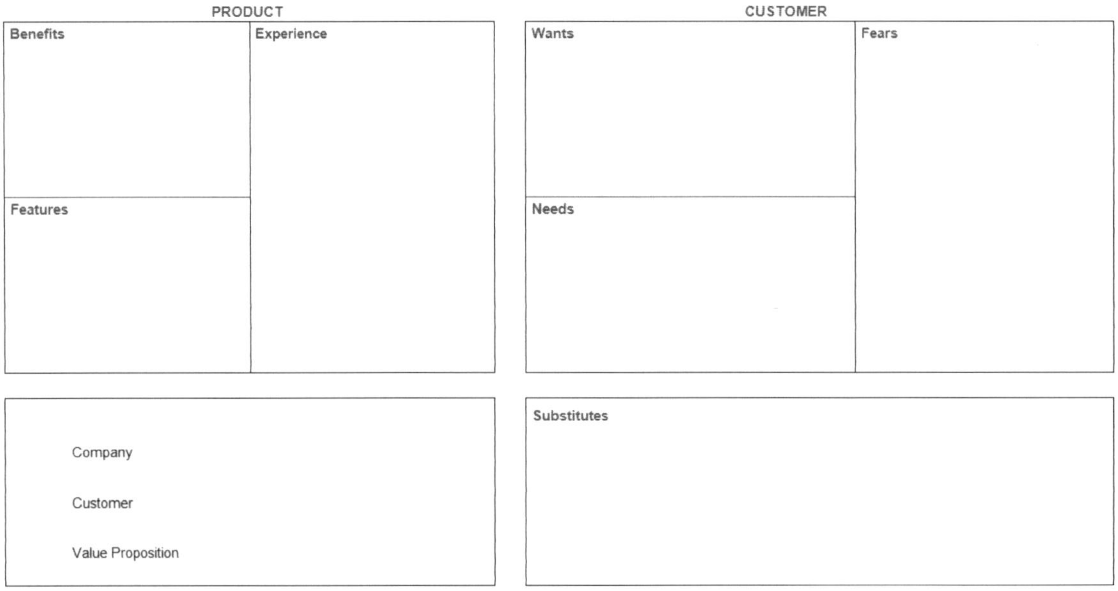

Figure 3. Value Proposition Canvas

Also during the workshops the participants could work in the templates, supported by the T-DoRe coaches and they also shared their own experiences in order to receive feedback from other participants.

The next tool they had to use was the business model canvas. This canvas is used to describe, design, challenge, and changes ina business model. It covers the following: Customer Segments, Customer Relationships, Channels, Value Proposition, Revenue Streams, Key Activities, Key resources, Key Partners and Cost Structure [7].

As a result of the itinerary, the 11 PhD participants developed a complete first version of their research project business case, including the value proposition canvas, the business model canvas, and also the validation board. This last element allowed the participants to formulate, analyze and validate hypothesis for the customer's problem and for the solutions to that problem in an iterative way. Through customer interviews, participants could collect direct feedback regarding the different hypothesis and then evolve it by discarding invalid hypothesis in order to progress in the definition of what is called minimum viable product (MVP).

3.1. Neodymium recovery from spent magnets business case

There is growing concern about the availability of rare earths as the demand for electronic devices and high technology products increases [8]. Amongst the rare earths, one of the most critical is neodymium (Nd), used in permanent magnets for applications such as turbines, hard disk drives, electric cars and electric storage systems. At present, toxicity and the complex energy intensive multi-stage processes limit the implementation of existing Nd recovery technology [9–11].

Electrochemical deposition of Nd metal is an interesting technology for the production of Nd from both, primary and secondary sources. However, it requires a suitable electrolyte media as the electrolysis of Nd is not possible in aqueous electrolytes due the insufficient electrochemical stability at the very negative potentials involved. Nd electrodeposition has been achieved in high temperature molten salts, but at temperature of 700 to 1000° C which also requires significant amounts of energy [10,11].

Ionic liquids (ILs) are purely ionic compounds, many with melting points below room temperature that are an interesting alternative media for Nd electrochemical recovery due to electrochemical stability. In recent years, the chemical separation and recovery of rare earths in ILs has been investigated, focusing primarily on those with imidazolium, betainium and ammonium cations [9,12–16] However, the electrochemical recovery of Nd in ILs has been scarcely investigated [16,17].

The objective of the PhD project associated to this business case is to develop a low temperature Nd electrodeposition process using ionic liquids as electrolytes as a greener and less energy intensive alternative to exiting processes.

3.1.1. Value proposition. The customer segment targeted in this business model is that of metal producers and especially those focused in precious metals although it does not exclude metal producers of common metals.

The value proposition canvas reflects customer requirements (needs, wants, fears) and the products/services that deliver value to that customer (benefits, experience, features). The following figure shows the value proposition canvas of this business case.

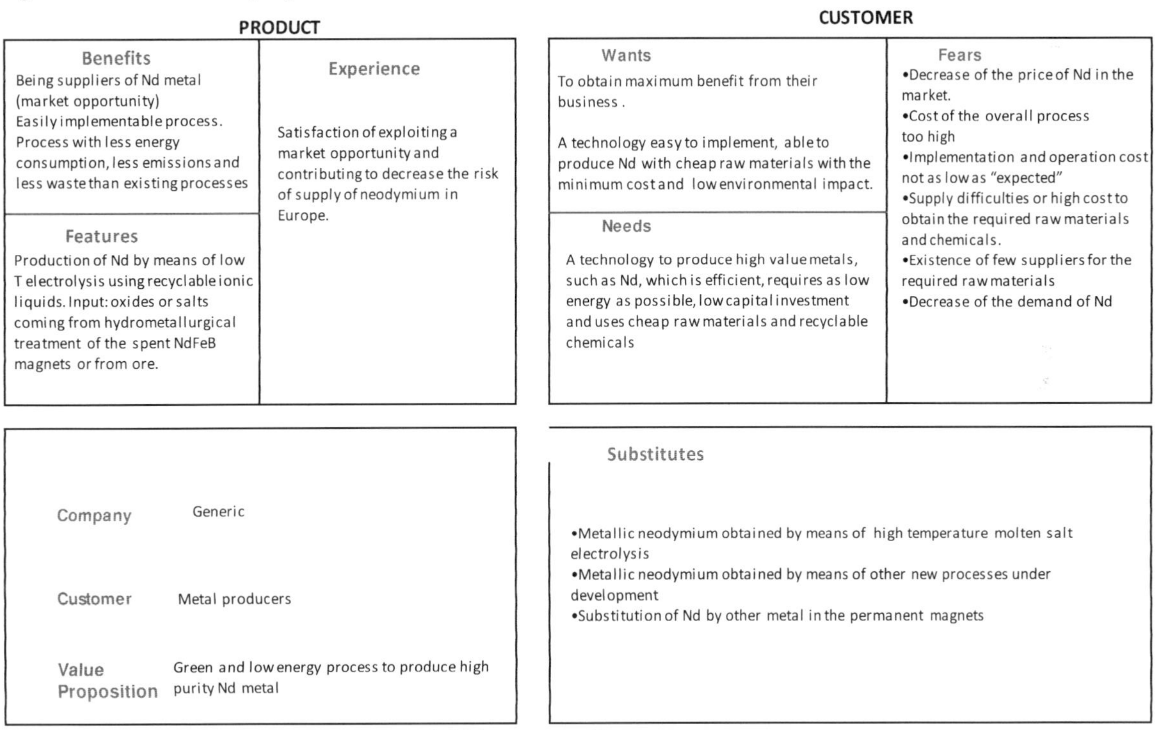

Figure 4. Nd recovery from spent magnets

Interviews with several customers allowed the researcher to obtain direct information of the needs and wants in order to understand them, prioritize and validate hypothesis.

The researcher identified aspects related to the potential market, and verified if the research was addressing the right issues.

3.1.2. Business Model Canvas. This canvas addresses the different elements that are related to the business model for a particular customer segment.

This canvas identifies the nine elements that explain the business model: Customer Segments, Customer Relationships, Channels, Value Proposition, Revenue Streams, Key Activities, Key resources, Key Partners and Cost Structure.

The customer segments identified are the metal producers, which is an important market segment because its large size, recyclers which already have hydrometallurgical plants and/or pyrometallurgical plants, rare earth metal and alloy producers, and electrolysis plants. This business model study has been carried out considering the customer segment of metal producers only because of its large size and because it has been the only one for which the researcher has found data about the size and other economic aspects.

The following figure shows the business model canvas of this business case.

<table>
<tr><td rowspan="2">Key Partners

1.Ionic liquids manufacturers
2. Engineering company (for plant desing and mechanical treatment optimisation)

3. Manufacturers of components for the electrolysis plant</td><td>Key Activities

Consulting/ provide our knowledge to implement the technology</td><td rowspan="2">Value Propositions

Green and Low T process for the production of high purity Nd metal by means of ionic liquid electrolysis (using recyclable ionic liquids)</td><td>Customer relationships

Consulting/ provide our knowledge to implement the technology</td><td rowspan="2">Customer Segments

Metal producers</td></tr>
<tr><td>Key Resources

1. Experts on the technology
2. Well equipped laboratories</td><td>Channels

1. Meetings & visits to their plant
2. Technical support and discussions by email or telephone</td></tr>
<tr><td colspan="3">Cost Structure

1. Office for the experts
2. Salary of the experts</td><td colspan="2">Revenue Streams
1. Payment for the license of our process
2. Our knowledge, consulting services to implement the technology
3. Agreement with Ionic Liquids manufacturer to obtain some commission from each tone of ionic liquid sold</td></tr>
</table>

Figure 5. Nd recovery from spent magnets Business Model canvas

The business model has been supported by calculations of revenues and costs considering some assumptions related to the customers, their problems and the proposed solutions. Although Technology Readiness Level (TRL) is still low, this exercise provided a wider vision in the PhD research.

Analyzing the complete value chain new questions and aspects to be treated arose, e.g. concerning the availability of raw materials for the proposed fabrication process, and therefore the researcher identified new elements that need to be analyzed as well as new stakeholders.

3.2. Impact of T-DoRe itinerary.

The impact of the methodology applied in the 11 business cases varied depending on the stage of the research and on the previous educational background and working experience of the researchers.

In general, several benefits have been detected, from the development of strategical and communication competences, to the analysis of new aspects in the project, covering the complete value chain of the project.

In the particular Nd recovery case, the researcher mentioned the positive impact of the methods and tools to incorporate in the analysis and development of the research, like the Value Proposition and Business Model canvases, and also other tools as the "Motorola" tool to prepare and afterwards analyze the visits to potential customers and to other stakeholders in the value chain.

In addition, the researcher highlighted the importance of having a good community learning environment during the itinerary. The community learning process made the itinerary more motivating and richer, as researchers had the opportunity to share reflections and questions, to learn from each other's experiences, as more contributions could enrich the business case of each participant. Last, the participation in a community learning process is a way to enlarge the network in the field of Raw Materials, and it can provide opportunities for future collaboration.

4. Conclusions

This paper explains the T-DoRe itinerary and how this methodology has helped the participants to widen their profile with traversal skills as teamwork, innovation and business orientation. Acquiring these skills has allowed participants to take ideas into action and to transform their PhD research knowledge into solutions (products or services) that can be transferred to the market. All this process has crystallised into each participants' Business Cases.

Some interesting results have been obtained. The business cases developed by candidates has shown that the creation of new holistic innovation champions through training of T-shaped professionals from Universities, RTOs and industry in the raw materials sector with knowledge in applied sciences, practice, and market innovations is possible.

The action oriented activity "customer interviews" has promoted the interaction of the participants with industry and stakeholders to understand real needs related with their research knowledge. This wouldn't have occurred without the T-DoRe itinerary.

The eleven participants have created their value proposition and business model with the same success that the Neodymium case. PhD participants have thought on their thesis as a product, process or service that a potential customer might need and likely buy.

The business model creation activity has helped them to better focus the research (in case of low TRL researches) and therefore define the future transferable research results. On the other hand, for higher TRL projects, the itinerary has helped to improve the value proposition for customers.

Future research value for low TRLs depends on the value propositions by these researches, and the advantages of focusing is clear. Technological value for companies depends on the business value that the company gets from that technology. For higher TRLs the approach is more straightforward as the focus is based on the short term. Anyway, technology is seen by companies as a means to an end, and not an end in itself.

About the types of business models developed, the results obtained have underlined the importance of the services as an innovation approach for the industry, as well as the fact that Scientific R&D consulting remains the most important approach for spin-outs.

Moreover, the fact that most of the business models have focused their market in the machine and equipment market underlines the idea that although raw materials industry is both product and process based, innovations in this sector are based equipment capable of creating and processing new materials and new products. Besides, the innovations proposed also emphasise the idea that opportunities for this industry are associated to the servitization and personalization of their business. Initially coined by Vandermerewe & Rada [18], servitization means the process of creating value in products and goods by adding services. Another trend in industry, and reflected in the business models, is personalised customisation. This trend is related with the creation of services to meet customer (or even individual) specific requirements with the implementation of highly personalized advanced services.

Finally, the business models developed have also highlighted the idea that although the innovations proposed vary depending on the position in the value chain, opportunities for the raw material industry to developed new business are business-to-business (B2B, in line with the fact that business-to-business (B2B) has seen an increase in private equity investments and investments in B2B start-ups.

The next step is to look for patterns among the business models defined by participants in order to have basic templates to help to think in how to bring the research results to the market with successful results and create innovative business.

Acknowledgements

We would like to thank the European Institute of Innovation & Technology (EIT) and the Knowledge Innovation Community of Raw Materials for their support in the development of T-DoRe project, as well as to all participating agents.

References

[1] Michigan State University T-SUMMIT 2016: Transformational approaches to creating t-shaped professionals *Conf.*, http://tsummit.org/t

[2] Kos B T-shaped skills in every area of your life AgileLeanLife https://agileleanlife.com/t-shaped-skills-every-area-life/

[3] Ruuska, Juha & Krawczyk, Piotr. 2013. Team Academy as Learning Living Lab. European Phenomena of Entrepreneurship Education and Development. University Industry Conference, Amsterdam

[4] Hanghøj K, 2015 EIT RawMaterials. Brussels https://eit.europa.eu/eit-community/eit-raw-materials

[5] Blank S, 2013 *The Four Steps to the Epiphany. Successful Strategies for Products that Win.* K&S Ranch; 2nd edition

[6] Osterwalder A, Pigneur Y, Bernarda G, Smith A. 2014 *Value Proposition Design: How to Create Products and Services Customers Want* John Wiley & Sons

[7] Osterwalder A, Pigneur Y, 2013 *Business Model Generation: A Handbook for Visionaries, Game Changers, and Challengers.* John Wiley & Sons

[8] Alonso E, Sherman A M, Wallington T J, Everson M P, Field F R, Roth R, Kirchain R E, Evaluating Rare Earth Element availability : A Case With Revolutionary Demand From Clean Technologies, Environ. Sci. Technol. 46 (2012) 3406–3414. doi:10.1021/es203518d.

[9] Villemin D, Didi M A, Extraction of rare earth and heavy metals, using ionic solvents as extraction medium (A review), Orient. J. Chem. 29 (2013) 1267–1284. doi:10.13005/ojc/290402.

[10] Mehra O K, Singh S, Sehra J C, Krishnan T S , Bose D K, Gupta C, Extractive Metallurgy of Rare Earths - Developmental Work at the Bhabha Atomic Research Centre, Mater. Sci. Forum. 30 (1988) 73–88. doi:10.4028/www.scientific.net/MSF.30.73.

[11] Tanaka M, Oki T, Koyama K, Narita H, Oishi T, Recycling of Rare Earths from Scrap, 1st ed., Elsevier B.V., 2013. doi:10.1016/B978-0-444-59536-2.00002-7.

[12] Yang F, Kubota F, Baba Y, Kamiya N, Goto M, Selective extraction and recovery of rare earth metals from phosphor powders in waste fluorescent lamps using an ionic liquid system, J. Hazard. Mater. 254-255 (2013) 79–88. doi:10.1016/j.jhazmat.2013.03.026.

[13] Vander Hoogerstraete T, Onghena B, Binnemans K, Homogeneous liquid-liquid extraction of rare earths with the betaine-betainium bis(trifluoromethylsulfonyl)imide ionic liquid system, Int. J. Mol. Sci. 14 (2013) 21353–21377. doi:10.3390/ijms141121353.

[14] Rout A, Kotlarska J, Dehaen W, Binnemans K, Liquid–liquid extraction of neodymium(iii) by dialkylphosphate ionic liquids from acidic medium: the importance of the ionic liquid cation, Phys. Chem. Chem. Phys. 15 (2013) 16533. doi:10.1039/c3cp52218k.

[15] Guocai T, Chapter 5. Application of Ionic Liquids in Extraction and Separation of Metals, in:

Green Solvents II Prop. Appl. Ionics Liq., 2012: pp. 119–153.

[16] Kondo H, Matsumiya M, Tsunashima K, Kodama S, Attempts to the electrodeposition of Nd from ionic liquids at elevated temperatures, Electrochim. Acta. 66 (2012) 313–319. doi:10.1016/j.electacta.2012.01.101.

[17] Kondo H, Matsumiya M, Tsunashima K, Kodama S, Investigation of Oxidation State of the Electrodeposited Neodymium Metal Related with the Water Content of Phosphonium Ionic Liquids, ECS Trans. 50 (2012) 529–538. doi:10.1149/05011.0529ecst.

[18] Vandermerwe S, Rada J, 1988 *Servitization of business: Adding value by adding services*, Eur. Manag. J., vol. 6, no. 4, pp. 314–324,

Real life experimental determination of platinum group metals content in automotive catalytic converters

I Yakoumis[1], A M Moschovi[1], I Giannopoulou[2], D Panias[2]

[1]Monolithos Catalysts and Recycling Ltd, 11476 Polygono, Athens, Greece
[2]Laboratory of Metallurgy, School of Mining and Metallurgical Engineering National Technical University of Athens, Athens, Greece
E-mail: yakoumis@monolithos-catalysts.gr

Abstract. The real life experimental protocol for the preparation of spent automobile catalyst samples for elemental analysis is thoroughly described in the following study. Collection, sorting and dismantling, homogenization and sample preparation for X-Ray fluorescence spectroscopy and Atomic Adsorption Spectroscopy combined with Inductive coupled plasma mass spectrometry are discussed in detail for both ceramic and metallic spent catalysts. The concentrations of Platinum Group Metals (PGMs) in spent catalytic converters are presented based on typical consignments of recycled converters (more than 45,000 pieces) from the Greek Market. The conclusions clearly denoted commercial metallic catalytic foil contains higher PGMs loading than ceramic honeycombs. On the other hand, the total PGMs loading in spent ceramic catalytic converters has been found higher than the corresponding value for the metallic ones.

1. Introduction

Catalytic converters are devices that control the exhaust emissions of an internal combustion engine by converting the exhaust toxic gasses into less toxic substances. Since 1994, when the European Directive 94/12/EEC (on ambient air quality) was implemented, all the motors produced or imported in EU are obliged to be fitted with catalytic converters that reduce emissions of carbon monoxide, hydrocarbons and nitrogen oxides below the legislative levels.

These converters are installed on all petrol-engine vehicles since 1993 and diesel-engine vehicles since 1997 in order to fight traffic-related pollution. The main part of the autocatalyst is an internal honeycomb structure, made either of ceramic or metallic material, coated by an alumina layer (called a ''wash coat'') that is enriched with PGMs. Today, the mostly used automotive catalytic converters are the "three-way converters" (TWC), which are comprised of Platinum (Pt) or Palladium (Pd) and Rhodium (Rh). Among these metals, Pt is the most active catalyst and is used both as an oxidation and as a reduction catalyst; Pd is used as an oxidation catalyst and Rh, as a reduction one. Such catalysts require careful engine design and management in order to ensure the engine operates in accordance to the stoichiometry. The reactions that take place in the converter are described in Figure 1.

$$2\,NO_x \xrightarrow{Rh} x\,O_2 + N_2$$

$$2\,CO + O_2 \xrightarrow{Pd} 2\,CO_2$$

$$C_xH_{2x+2} + \frac{(3x+1)}{2}O_2 \xrightarrow{Pt} x\,CO_2 + (x+1)\,H_2O.$$

Figure 1. Catalytic reactions of nitrogen oxides, oxidation of carbon monoxide and oxidation of unburnt hydrocarbons in automotive catalysts.[1]

A large reactive surface area is required to convert the high speed exhaust gases. This is accomplished by the surface enlarging wash coat. The surface of a conventional substrate is approximately 3m^2 at 1dm^3 and is multiplied to thousands of m^2 using the wash coat. Commercial automotive catalysts are separated in two different categories, depending on the substrate which is implemented. The most common ones are the catalytic converters using ceramic substrate i.e. 90%, while the 10% of the used converters have metallic substrate.[2] The difference in the structure of the converters is described in Figure 2, while physical properties comparison is described in Table 1.

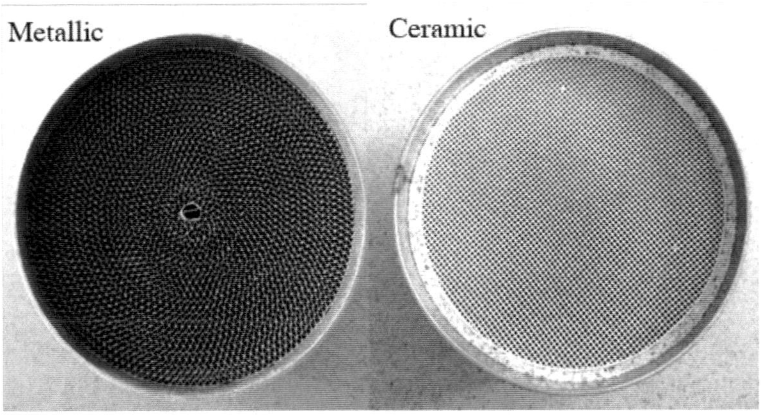

Figure 2. Structural differences between Metallic and Ceramic monolithic substrate.

Metallic converters have demonstrated advantages over the physical properties of ceramic converters, such as lower wall thickness and lower thermal resistance, which enhance flow rate through the metallic network and horsepower. Moreover, metallic infrastructures are easier to pack, less fragile and more robust and tolerant in vibrations. On the other hand, wash coat adhesion issues on metallic parts arise in heating cooling cycles.

Table 1. Physical properties of ceramic and metallic converters.[3]

	Ceramic converter	Metallic converter
Cell density(#/cm^2)	62.00	62.00
Wall Thickness(mm)	0.16	0.05
Solid Fraction	0.24	0.09
Material density (g/cm^3)	1.70	7.10
Bulk Density(g/cm^3)	0.41	0.63
Material Porosity (%)	35.00	0.00
Heat Capacity (kJ/L°C)	0.43	0.32
Thermal expansion (Rel)	1.00	17.00
Thermal conductivity (Rel)	1.00	20.00

A mixture of silica and alumina is often used wash coat on the honeycomb substrate. The wash coat forms and irregular surface which has a far better surface area than the flat substrate, providing the autocatalyst with more place for active precious metal sites. In the case of wash coat usage, the catalyst is added to the wash coat before application to the core. After years of exposure in harsh and severe conditions of a catalytic reaction process, spent catalyst have no longer homogeneous catalyst distribution. By the end of their life, either they have accumulated many different contaminants of various densities, organic compounds, oils and water or they are broken. Examples of ceramic and metallic de-activated catalytic convertors are shown in Figure 3. Thus, once they are recycled, appropriate sampling procedure should be established in order to obtain a homogeneous representative sample for qualitative and quantitative analysis.

Although there are a broad variety of possible combinations of PGMs in the catalytic converters, bibliographic data suggest that the total content of these metals is up to 2,000 ppm in the ceramic substrate or in some cases 10^{-3}-10^{-1}%.[4,5] In other bibliographic references it is mentioned that an automobile catalytic converter contains typically 0.08% Pt, 0.04% Pd and 0.005-0.007% Rh supported on a base.[6,7] Dong *et al.* [8] also refer that the total PGMs loading fluctuates between 0.1-0.5%. Moreover, Kuczynski *et al.* [9] suggest that automobile catalysts can contain up to 30 tr oz of PGMs per short ton of catalyst (1,000 ppm). Kim *et al.* [10] refer that an autocatalyst essentially comprises a refractory oxide support, on which two or more precious metals are dispersed in very low concentrations (0.1-0.3 wt% of the monolith). Ravindra *et al.* [11], generally, mention that a vehicle exhaust catalyst contains around 1-3 g of PGMs corresponding to approximately 1.8 mg PGMs/cm^3 of the catalyst. Finally, Barefoot *et al.* [12] refers that platinum concentrations, which are generally larger than those of the other precious metals, have ranged from 300 to 1,000 μg/g, palladium from 200 to 800 μg/g and rhodium from 50 to 100 μg/g. On the other side, Jimenez de Aberasturi *et al.* [13] and Fontas *et al.* [14] suggest that the total content is always lower than 1,000 ppm.

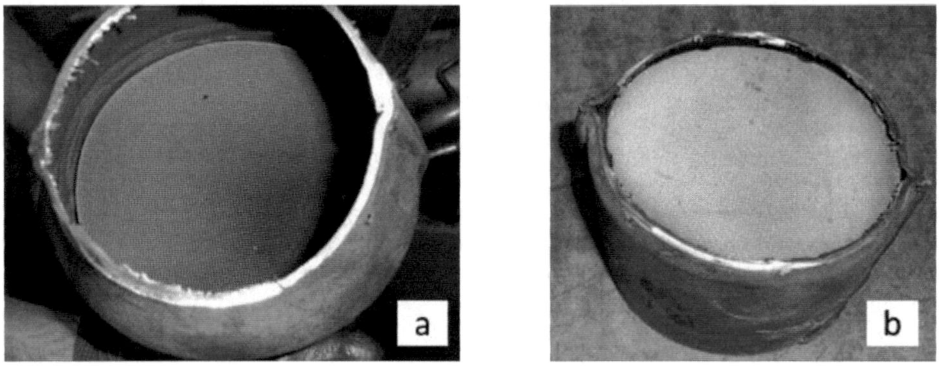

Figure 3. Deactivated (a) Metallic and (b) Ceramic autocatalysts.

2. Recycling process steps

Recycling of the spent automotive catalysts targets the extraction of the contained precious metals, contributing significantly towards safeguarding of the natural resources and ensuring the minimization of the environmental pollution, related both to the catalysts disposal and PGMs production. Additionally, the high value of PGMs in spent catalysts justifies their recycling. Nowadays, the state of the art technology is pyrometallurgy. Spent chemical catalysts are considered non-hazardous wastes according to European Regulation EC No 2150/2002. Effective recycling requires a well-tuned recycling chain, consisting of different specialized stakeholders. The whole process is designed to enable the separation and recovery of the precious metals from automotive catalysts, with the highest possible yield and efficacy. The main parts of recycling procedure are described in Figure 4 and discussed below.

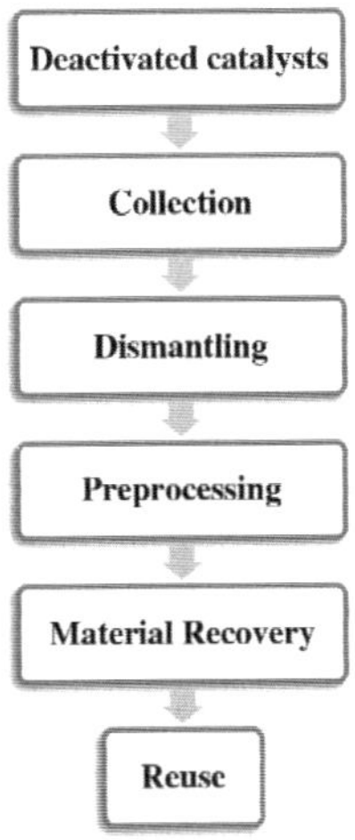

Figure 4. Recycling process steps of spent catalysts.

2.1 Collection and classification of the converters from the scrap yards or car service workshops.
This step is performed throughout Europe from numerous small-sized companies and natural entities. Appropriate classification is the foundation for ensuring that the collection, transportation, storage and treatment of spent autocatalysts are carried out in a manner that provide protection for the environment and human health and in compliance with legal requirements (Figure 5).

Figure 5. Collection and classification of spent automobile catalysts.

2.2 Sorting and dismantling the converters.
The precious-metal-bearing carrier material and the non-precious-metal-bearing steel casing are separated (De-canning of ceramic converters / Shredding of metallic converters). Inlet and outlet exhaust extension tubes are cut (Figure 6) and main bodies of the catalysts proceed to the sorting step (Figure

6). This process is mainly done some form of hydraulic shear. The PGMs content in each converter is highly dependent on:

- the power of the engine of the car,
- the year of manufacturing of the car,
- the fuel that the car uses (diesel or gasoline) and
- if the converter is the genuine part or a replacement part (Original Equipment Standard or Aftermarket).

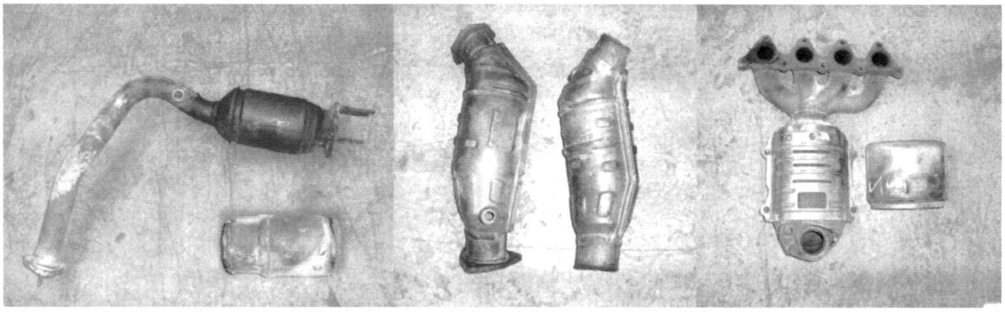

Figure 6. Inlet and outlet exhaust tubes are removed from the main body of the catalysts during the shredding procedure.

After the sorting, the converters are de-canned, i.e. the metallic canister and the wool (fibre mat) are removed from the ceramic catalyst using mechanical process which is usually called guillotine or modified alligator shear (Figure 7). Most damage occurs during the dismantling process. The steel casings are removed by an automatic cutting machine. This may also damage the catalyst inside, while further losses may occur through handling and transit.

Figure 7. (a) Automatic cutting machine for the ceramic catalyst decanning (b) empty case of ceramic catalyst and (c) crashed ceramic honeycomb.

2.3. Homogenization, sampling and assaying.

The ceramic monoliths (or the metallic converters) are driven to a pre-crusher in order to downsize their dimensions and to a mill in order to extract a fine homogeneous powder, which contains the PGMs. Metallic foils are mechanically separated from the wash coat, which contains the PGMs.

Different minerals behave differently during pulverization – most (brittle) minerals will easily break down to small particles while some will just change their shape if proper sample preparation methods are not used. Due to the fact that the recovery of PGMs requires sophisticated large-scaled metallurgical processes, the most common approach is to extract samples from each consignment in order to determine the PGMs content of each lot, in every stage of the procedure. The subsequent sampling process is used to collect a small representative sample of material on the basis of which a valuation is made for several tonnes of material.

2.4. Recovery and refining facilities

Recovery and refining is mainly carried out by large infrastructures. Umicore, at its integrated smelter-refinery in Antwerp, Belgium, currently recovers and supplies back to the market via its main process route seven precious metals (amongst them Platinum, Palladium and Rhodium). Johnson Matthey, at its refineries in Brimsdown and Royston, UK, recovers PGMs. Overseas, Multimetco Inc. (Anniston, Alabama, USA) and Techemet LP (Pasadena, Texas, USA) are specialized in refining PGMs from automotive catalysts. At this stage, the recovery of PGMs, their separation and purification are accomplished. The main process technologies applied today for the PGMs recovery from the spent automotive catalytic converters are classified in three groups: (a) hydrometallurgical, (b) pyrometallurgical and (c) volatilization.

The industrial recovery rates of PGMs are extremely high, i.e. for Pt and Pd is 98%, while the recovery for Rh is 87.5%. The cost of the refining process is more or less 4,000€/ton of ceramic monoliths or 2,500€/ton of metallic converters. The refining cost represents 6.5% of the actual value of the recovered PGMs for ceramic monoliths and 13.6% of the actual value of the recovered PGMs for metallic catalytic converters. The total value of recovered PGMs, based on economic figures of July 2017, is estimated at 73,000€ per ton of ceramic monoliths. The metals prices are based on Johnson Matthey Precious Metal Management sales prices for 16 October 2017. More specifically the prices were: Pt: 28,436€/kg, Pd: 30,343€/kg and Rh: 39,539€/kg.

3. Experimental

To accurately determine the amount of precious metal present in the monoliths, suitable samples for laboratory analysis were prepared. The goal of the experimental protocol is to obtain a sample that is a precisely representative of the overall material, as homogenous as possible of the recoverable metals within the lot. Deficient sampling and sample preparation can have high financial impact. Moreover, a constant protocol ought to be firmly followed in order for the measurements to be reproducible and comparable.

Measurements took place in ceramic and metallic batches of commercial spent catalysts. Twenty nine ceramic batches consisting roughly of 1,400 pieces per batch and nine metallic batches consisting in average of 450 pieces per batch were converted into homogeneous mass. The sampling and homogenization processes are described in Figures 8a for ceramic honeycombs and Figure 8b for metallic foils.

One batch, consisting of approximately 1,400 spent catalysts, was originally de-canned. The total mass of each batch for the ceramic honeycombs was roughly 1,047kg, which in average is 0.7374kg per honeycomb with particle size 10cm.

Each batch was introduced to the ball mill (Figure 9a) in order to obtain particle size 2mm. 90% of the milled powder was send for recycling/ storage and the rest 10%, (100kg) remained in the ball mill and grounded again, in order to obtain particle size 1mm. 90% of the milled powder was removed and sent for recycling/storage and 10% (10kg) proceeded in the analytical mill (Figure 9c) to grounded to 160μm.

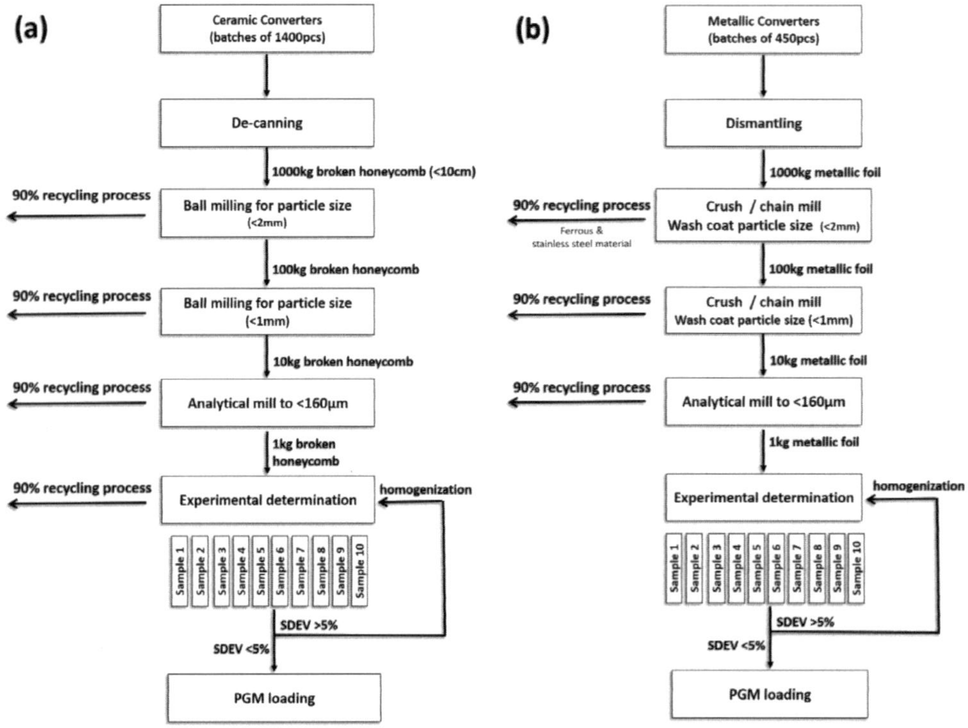

Figure 8. Sampling and homogenization process for (a) ceramic and (b) metallic catalytic converters.

Finally, 90% of the powder was sent for recycling/storage and the rest 1kg was used for elemental analysis. Ceramic honeycomb final step powder was collected so as to prepare one set of 10 samples, each one of 10g. Each sample was used for elemental XRF analysis. In case where the standard deviation of the identified values of the set was greater than 5%, then samples were mixed again and proceeded with homogenization. On the other hand, if standard deviation was lower than 5%, the experimental values were reliable, revealing that the batch was homogeneous.

Similar procedure was also followed for the treatment of metallic catalysts. Nine batches, consisting in average of 450 catalysts per batch were used for analysis. The metallic catalysts were initially crushed, where ferrous and stainless steel metallic parts were removed. The average mass of the metallic foil was 0.2143kg per catalyst. The metallic foils were introduced into the crush mill (Figure 9). 90% of the metallic foil was sent for storage and recycling and the rest 10% remained in the mill to be grounded again in order to obtain particle size 1mm. 90% of the milled foil with particle size 1mm was sent for storage/recycling and the rest 10% proceeded to the analytical mill. Analytical mill downsized the particle size to 160μm. Finally, 10% of the fine catalyst powder was used to prepare one set of 10 samples in order to verify the PGM loading on ceramic catalyst. In case where the standard deviation of the identified values of the set was higher 5%, then the samples were mixed again and proceeded with homogenization. On the other hand, if standard deviation was lower 5%, the experimental values were reliable, revealing that the batch was prepared homogeneously. The size of the particles was determined between the mill crushing steps using a series of stainless steel sieves with appropriate Mesh porosity.

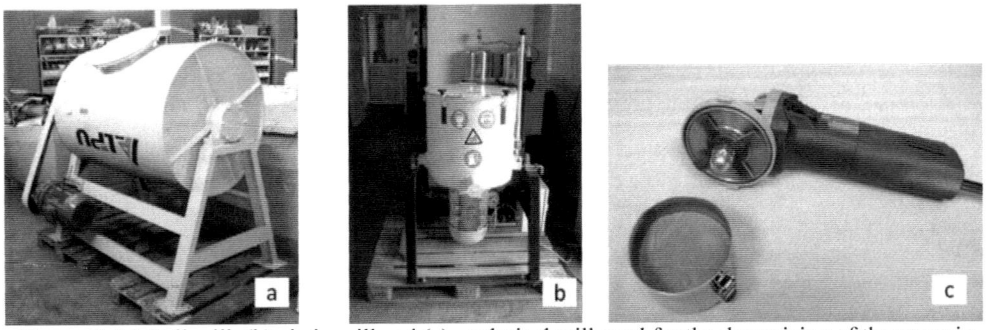

Figure 9. (a) Ball mill, (b) chain mill and (c) analytical mill used for the downsizing of the ceramic monolith particles before the analysis.

The content of spent catalysts in Pt, Pd and Rh was determined with X-Ray fluorescence spectroscopy (XRF) on solid samples. Since, small variations in analytical determinations impact critically on financial performance, XRF analysis was used for determination of noble metals. The XRF analyzer provides a high level of performance when measuring metals such as Pt, Pd, Rh in catalytic converters loading and in recovered metals from spent catalysts. While benefiting from the convenience of fast analysis, it is also of high confidence obtaining results that help assure high level of accuracy, repeatability and precision of almost 1%. The accuracy of the method is verified by ICP-MS technique.[15-18]The content of the other metals existing in the ceramic monolith and the wash coat were determined with a combination of the Atomic Adsorption Spectroscopy (ASS) and the Inductive Couple Plasma Mass Spectrometry (ICP-MS), using liquid samples produced after fusion of solid samples with LiB_4O_7 & KNO_3. Before XRF analysis and fusion procedure, the solid samples were dried overnight at 110°C for the adsorbed water to be removed.

X-Ray Fluorescence Spectroscopy

The absorption of X-Ray radiation is element and concentration dependent thus the radiation is absorbed more strongly by elements with higher molar mass. Moreover, the higher the concentration of a metal the more is the radiation absorbed.[19] As a result, X-Ray techniques are appropriate for qualitative and quantitative analysis of noble metals.[20] Additionally, in very short time, reliable results are available. Chemical analysis of autocatalyst samples was performed using a SPECTRO Xepos spectrophotometer that is established in the Laboratory of Metallurgy of the National Technical University of Athens.

Combined technique of XRF, AAS and ICP-MS

The laboratory elemental analysis of the spent autocatalysts samples was demonstrated with a combined technique that was based on the XRF method for the determination of PGMs and on the methods of Atomic Adsorption Spectroscopy and Inductive Couple Plasma Mass Spectrometry for the determination of the other elements existing in samples. For AAS method, a Perkin Elmer 2100 Spectrometer was used, while for ICP-MS method, a Thermo Scientific X-Series II Spectrometer was used. Both instruments are established in the Laboratory of Metallurgy of the National Technical University of Athens.

4. Results - PGMs content in spent catalytic converters

For the most accurate and reliable determination of the noble metal concentration on ceramic honeycombs and metallic foils, we chose to investigate a large number of auto-catalysts in order to obtain. the highest possible homogenization and representative values. More than 40,000 ceramic catalytic converters and more than 4,000 metallic catalysts were collected and treated for the elemental analysis.

The precious metal loading in each batch of ceramic and metallic catalysts was determined by X-Ray Fluorescence spectroscopy and the results are described in Table 2 for the ceramic honeycombs and in Table 3 for the metallic foils. Commercial data of an actual recycled quantity of more than 40,000 catalytic converters demonstrate that the mean concentration of PGMs in spent ceramic catalytic converters is 2,596 ppm or 0.26% w/w for the case of ceramic catalysts (Table 2). The mean concentration is 200% higher for the case of metallic catalytic foil, i.e. 7,872 ppm or 0.79% (Table 3).

Table 2. PGMs concentration of twentynine typical consignments of spent ceramic catalytic converters recycled from the Greek market.

	weight of ceramic honeycomb(kg)	**Number of Ceramic converters**	**Weight of ceramic monoliths (kg)**	**Pt (ppm)**	**Pd (ppm)**	**Rh (ppm)**	**Total PGMs (ppm)**
1	0.758	1,600	1,212.27	724	1,633	302	2,659
2	0.740	1,597	1,181.75	731	1,521	255	2,507
3	0.739	1,526	1,127.28	760	1,620	305	2,685
4	0.737	1,374	1,012.12	673	1,734	315	2,722
5	0.734	1,377	1,010.76	710	1,471	295	2,476
6	0.739	1,516	1,120.41	734	1,522	285	2,541
7	0.729	1,383	1,008.89	705	1,606	283	2,594
8	0.720	1,369	985.09	749	1,573	308	2,630
9	0.700	1,476	1,032.44	630	1,760	299	2,689
10	0.706	1,496	1,056.03	661	1,701	343	2,705
11	0.736	1,360	1,001.32	720	1,533	301	2,554
12	0.780	1,270	990.52	811	1,426	294	2,531
13	0.756	1,335	1,008.90	802	1,541	284	2,627
14	0.791	1,280	1,012.54	830	1,505	312	2,647
15	0.770	1,304	1,005.00	821	1,360	302	2,483
16	0.788	1,336	1,053.54	885	1,302	293	2,480
17	0.775	1,287	997.07	819	1,373	287	2,479
18	0.748	1,329	994.29	660	1,702	295	2,657
19	0.729	1,369	997.33	693	1,572	290	2,555
20	0.756	1,368	1,034.25	656	1,611	286	2,553
21	0.725	1,382	1,001.54	625	1,780	312	2,717
22	0.748	1,358	1,015.73	705	1,531	276	2,512
23	0.745	1,351	1,007.44	801	1,511	285	2,597
24	0.696	1,611	1,120.87	711	1,818	298	2,827
25	0.714	1,405	1,003.94	743	1,884	308	2,935
26	0.687	1,459	1,001.85	673	1,784	306	2,763
27	0.727	1,386	1,007.01	625	1,532	264	2,421
28	0.696	1,959	1,362.74	634	1,553	289	2,476
29	0.718	1,401	1,005.94	601	1,423	248	2,272
average	**0.737**			**720.414**	**1,582.138**	**293.793**	**2,596**

As it is already mentioned, a typical consignment of spent ceramic automotive catalytic converters contains 2,596 ppm of Platinum Group Metals (Platinum, Palladium and Rhodium) in the ceramic substrate (or 1.913g PGMs per catalytic converter, average weight of ceramic converter = 0.737 kg). The mass ratio of the three precious metals is: 2.7/ 6.2/ 1.1 Pt/ Pd/ Rh. This rate is highly dependent on the commercial price of each metal.

Table 3. PGMs concentration of nine typical consignments of spent metallic catalytic foils recycled from the Greek market.

	Weight of metallic foil (kg)	Metallic converters	Weight of metallic foils (kg)	Pt (ppm)	Pd (ppm)	Rh (ppm)	Total PGMs (ppm)
1	0.206	484	99.61	1,619	5,877	757	8,253
2	0.233	482	112.234	1,130	6,442	692	8,264
3	0.214	445	95.344	1,766	5,411	619	7,796
4	0.210	423	88.621	1,503	6,334	748	8,585
5	0.208	412	85.685	1,608	4,935	625	7,168
6	0.242	447	107.966	1,851	4,718	480	7,049
7	0.209	438	91.338	1,555	6,306	654	8,515
8	0.223	439	98.098	1,242	5,112	577	6,931
9	0.184	529	97.59	1,069	6,587	626	8,282
average	**0.214**			**1,482.55**	**5,747**	**642**	**7,872**

Quantitative analysis of nine batches of metallic foils showed that the average PGM loading was 7,872ppm on the foil (or 1.68gr PGM per catalysts, with average foil weight 0,214kg). The mass ratio of the noble metals is 1.02/ 3.9/ 0.5 Pt/ Pd/ Rh and the average PGM loading is 1,482ppm for Pt, 5,747ppm for Pd and 642ppm for Rh.

Analytical measurements of PGMs and the other elements of spent catalytic converters were performed on one single batch and the results are presented in Table 4. The total amount of noble metals is 0.2375% w/w or 2,375 ppm (Table 4). Therefore, the laboratory analysis confirms the aforementioned commercial catalysts measurements. Furthermore, the poisoning of the wash coat of the ceramic honeycomb is clearly listed in Table 4.

The studied ceramic-based catalyst contains up to 40% SiO_2, up to 10% MgO and up to 5% Ce, Zr and Ti and other mixtures introduced from gasoline and oil additives during operation. The identified values are in accordance with spent catalysts studied in the literature. Several oxides such as Na_2O and P_2O_5 and Fe_2O_3 were also determined on the structure of the spent catalyst as poisoning substances in particular high concentration.

Table 4. Laboratory chemical analysis of typical spent catalytic converters ceramic honeycomb.

Oxide/Metal	w/w	(ppm)
Ceramic Monolith - Cordierite		
MgO	8.51%	85,100
Al_2O_3	37.15%	371,500
SiO_2	35.33%	353,300
Poisoning (Lubricants - Fuels)		
Na_2O	3.07%	30,700
P_2O_5	1.81%	18,100
CaO	1.16%	11,600
Cr_2O_3	0.80%	8,000

Fe_2O_3	2.44%	24,400
ZnO	0.33%	3,300
BaO	0.77%	7,700
S	0.44%	4,400
C	0.43%	4,300
Catalyst Support (Wash coat)		
TiO_2	0.29%	25,900
ZrO_2	4.55%	45,500
La_2O_3	0.29%	2,900
CeO_2	3.51%	35,100
Catalyst (Noble Metals)		
Pd	0.1364%	1,364
Pt	0.0754%	754
Rh	0.0257%	257
PGMs sum	**0.2375%**	

In both cases, the PGMs concentration in spent automotive catalytic converters is significantly higher than the PGMs content in primary ores (on average 10-20 ppm). The high intrinsic metal values make recycling attractive from an economic point of view and, due to the much higher concentration compared to mined ores, it also helps to reduce the environmental burden of metal supply significantly, especially with the respect to the impact of climate.

5. Conclusions

The recycling of platinum group metals from spent automotive catalysts is becoming increasingly important. Thus, the procedures related to the evaluation of the spent catalyst and the recycled metals loading ought to be designed in detail. This study has determined that particular steps should be constantly followed in sampling procedure in order to obtain representative, reproducible and high quality analytical data for spent catalysts loading.

In this work more than 40,000 spent ceramic catalytic converters have been collected and pre-processed in order to extract homogeneous samples of ceramic honeycomb. The average content of ceramic honeycomb has been found to 2,596ppm or 1.913 gr of PGMs per catalyst. In the same time more than 4,000 spent metallic catalysts have been collected and pre-processed in order to extract honeycomb metallic foil. The average PGM content of metallic honeycomb has been found to be 7,874ppm (or 1.68gr of PGM per spent metallic converter). The PGM loading in ceramic honeycomb was also verified with laboratory analysis based on a combination of XRF, AAS and ICP-MS methods, where less than 1% deviation was noticed for the combined technique than XRF. In conclusion, XRF analysis provides an accurate and quick tool to determine PGM loading reducing cost, since the use of expensive lab wet chemistry reagents is avoided. Further, the analyses are quick and straightforward without the need of being contacted by highly educated and experienced analytical professionals.

6. References

[1] Ertl G Knözinger H Schüth Veitkamp J **2008**, Handbook of Heterogeneous Catalysis ISBN: 978-3-527-31241-2

[2] Hensel C Konieczny R Brück R **2000**, Recycling Technology for Metallic Substrates: A Closed Cycle SAE International

[3] Vaneman G L **1991**, Vaneman, G. L., Comparison of Metal Foil and Ceramic Monolith Automotive Catalytic Converters *Stud. Surf. Sci. Cat.* 537

[4] Hagelüken B C **2012**, Recycling the Platinum Group Metals: A European Perspective *Plat. Met. Rev.* **56** (1) 29

[5] Saternus M Fornalczyk A **2013**, Possible ways of refining precious group metals (PGM) obtained from recycling of the used auto catalytic converters *Metalurgija* **52** (2) 267
[6] Rao C R M Reddi G S **2000**, Platinum group metals (PGM); Occurrence, Use and Recent Trends In Their Determination. *TrAC Tr. Anal. Chem.* **19** (9) 565
[7] Hoffmann J E **1988**, Recovery of Platinum-Group Metals from Gabbroic Rocks Metals from Auto Catalysts *JOM* **40** (6) 40
[8] Dong H Zhao J Chen J Wu Y Li B **2015** Recovery of platinum group metals from spent catalysts: A review *Int. J Miner. Proces.* 145 108
[9] Kuczynski R J Atkisnson G B Walters L A **1992** High temperature cyanide leaching of platinum group metals from automobile catalysts process development unit report of investigations 9428 US Bureau of mines
[10] Kim C H Woo S I Jeon S H **2000**, Recovery of Platinum-Group Metals from Recycled Automotive Catalytic Converters by Carbochlorination. *Ind. & Eng. Chem. Res.* **39** (5) 1185
[11] Ravindra K Bencs L Van Grieken R **2004**, Platinum group elements in the environment and their health risk. *Sci. The Tot. Envi.* **318** (1) 1
[12] Barefoot R R **1997**, Determination of Platinum at Trace Levels in Environmental and Biological Materials *Envir. Sci. Techn.* **31** (2) 309
[13] Jimenez de Aberasturi D Pinedo R Ruiz de Larramendi I; Ruiz de Larramendi J I Rojo T **2011**, Recovery by hydrometallurgical extraction of the platinum-group metals from car catalytic converters *Min. Engin. 24* (6) 505
[14] Fontàs C Salvadó V Hidalgo M **2002**, Separation and Concentration of Pd, Pt, and Rh from Automotive Catalytic Converters by Combining Two Hollow-Fiber Liquid Membrane Systems. *Ind. & Engin. Chem. Res. 41* (6) 1616
[15] Sun PP Lee MS **2013,** Recovery of Platinum from Chloride Leaching Solution of Spent Catalysts by Solvent Extraction *Mater. Trans.* **54** (1) **74**
[16] Masuda C Yonezu K Watanabe K Tokoyama T **2013** Recovery of Platinum From Spent Reforming Catalyst by Acid Leaching and Coprecipitation *Proc. Ear. Plan. Sci.* **6** 435
[17] Jafarifar D Daryanavard MR Sheibani S **2005** Ultra fast microwave-assisted leaching for recovery of platinum from spent catalyst *Hydrometal* **78** (3-4) **156**
[18] Smekens A Behets M Kazandjian P Van Grieken R **2007** Determination of platinum, palladium, and rhodium in automotive catalysts using high-energy secondary target X-ray **79** (16) **6383**
[19] Hagelüken C Jongen L Rieke T Klümper T **2010**, Process and Apparatus for identifying autocatalysts. Appl, No.1 12/872,478 Pub No US 2011/0075801 A1
[20] Renner H Schlamp G Kleinwächter I Drost E Lüschow H M Tews P Panster P Diehl M Lang J Kreuzer T Knödler A Starz K A Dermann K Rothaut J Drieselmann R Peter C Schiele, R **2000**, Platinum Group Metals and Compounds In *Ullmann's Encyclopedia of Industrial Chemistry*, Wiley-VCH Verlag GmbH & Co. KGaA

High temperature coatings from post processing Fe-based chips and Ni-based alloys as a solution for critical raw materials

T Dudziak[1*], A Olbrycht[2], A Polkowska[1], L Boron[1], P Skierski[1], A Wypych[3], A Ambroziak[4] and A Krezel[5]

[1]Foundry Research Institute, Zakopiańska 73, 30-418 Kraków, Poland
[2]Institute of Precision Mechanics, ul. Duchnicka 3, 01-796 Warsaw, Poland
[3]Wrocław University Science and Technology, Wybrzeże Wyspiańskiego 27 50-370 Wrocław, Poland
[4]Poznań University Science and Technology, Plac Marii Skłodowskiej-Curie 5 60-965 Poznań, Poland
[5]Tadeusz Kościuszko Univeristy of Technology, Warszawska 24, 31-155 Kraków, Poland

* corresponding author tomasz.dudziak@iod.krakow.pl

Abstract

Due to shortage of natural resources worldwide, it is a need to develop innovative technologies, to save natural resources and secure Critical Raw Materials (CRM). On the other hand, these new technologies should move forward materials engineering in order to develop better materials for extreme conditions. One way to develop new materials is to use post processing chips of austenitic steels (i.e. 304L stainless steel: 18/10 Cr/Ni) and other materials such as Ni-based alloy with high Cr content. In this work, the results of the preliminary study on the High Velocity Oxy Fuel (HVOF) coatings developed from 304L stainless steel chips and Haynes® 282® Ni- based alloys are shown. The study obeys development of the powder for HVOF technology. The produced coatings were exposed at high temperature at 500 and 700 °C for 100 and 300 hours respectively to assess corrosion behaviour.

Keywords: recycling, coatings, high temperature, steel, Ni-based alloy

1. Introduction

Corrosion represents a significant cost and technical challenges for developed economies affecting mostly constructions made of plain carbon steels or low alloyed steels: the materials often used in power plant, automotive, and other sectors of industry. Under ambient conditions, these quickly undergo corrosion processes leading to significant material loss that contribute strongly in economic losses providing high costs of repair or replacement [1]. Cost of boiler tube repair includes as well cost return to service and cost of lost production are generally in range of $10,000 – $100,000 depending of damage and difficulty accessing to the failure location, average costs reach around $10,000 per day. Overall, industrial boilers is one of the prominent features in fossil fuel fired power plants. The boilers constructed from low alloyed steel (T22) show failures mostly due to fuel ash corrosion, overheating, hydrogen attack, carburization and decarburization, corrosion fatigue cracking, stress corrosion cracking, caustic embrittlement, erosion, and steam oxidation. The failures are closely linked with wall thinning effect due to corrosion called metal loss effect [2, 3]. The majority of forced outages of power boilers are due to premature failure of boiler components [4, 5]. Boiler tube failures are the main cause of forced outages of power generating units. In the United States (US), the cost penalty is estimated to be in excess of 5 billion dollars a year in replacement power charges and maintenance costs [6]. The contribution of total tube failures can be grouped as: furnace water wall tubing 40%, super heater tubes 30%, re-heater (RH) tubes 15%, economizer tubes 10% and burner tubes 5% [7].Hence, it can be assumed that around 3.14% GDP of the European Union (EU) may end up with direct costs of corrosion reaching 250 billion of € [8,9]. The mentioned costs showing, high impact of corrosion in different fields for GDP of EU, therefore it is extremely important to perform more scientific research in order to reduce provided corrosion costs. The highest, most accelerated corrosion degradation is observed at high temperatures as mentioned earlier: corrosion as a process is activated by temperature; as shown in **Figure 1.** The highest corrosion rate (metal loss) is met in low alloyed steels such as 15Mo3, T22, and T23 (blue stripe in **Figure 1**); the steels are unable to promote development of a stable, thin protective oxide scale. On the other hand, low alloyed steels cannot be replaced in full in power industry by expensive highly alloyed steels and Ni-based alloys, because low alloyed steels must be used in boiler walls due to much higher coefficient of thermal conductivity (CCT) and lower coefficient of thermal expansion (CTE) than highly alloyed steels (austenitic steels) and Ni-based alloys [10].

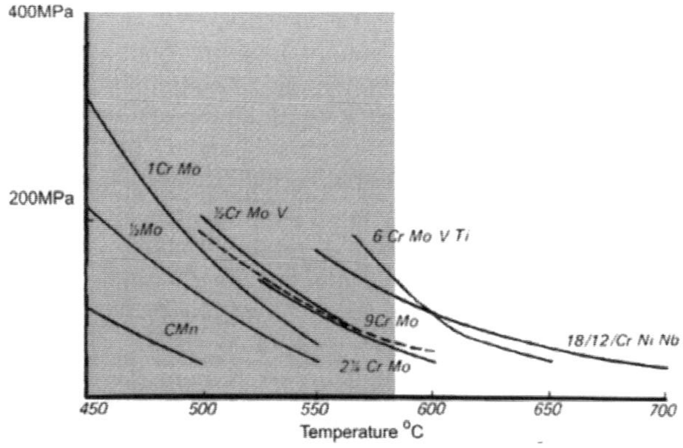

Figure 1 Material limits at high temperature for power plant industry

In order to reduce corrosion costs in low alloyed steels in power plant industry and in other sectors where such steels are used, increase efficiency due to higher operating temperature (**Figure 2**), enhance performance, reduce metal loss, a protective coating with a tailored chemical composition ensuring development of stable, thin, and adherent oxide scale have to be applied.

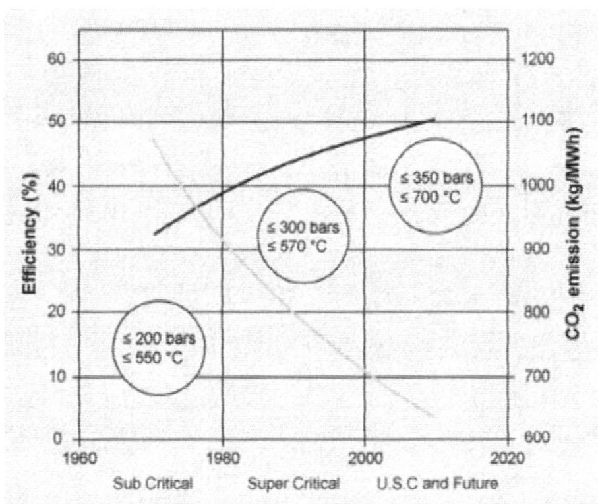

Figure 2 Efficiency and CO_2 emission in function of temperature and technology

Furthermore in nowadays economically and strategically important for the European economy, is secure of Critical Raw Materials (CRM). The materials such as Antimony, Berylium, Borates, Chromium, Cobalt, Cooking Coal, Fluorospar, Gallium, Magnesite, Magnesium, Natural Graphite, Niobium, Phosphate Rock, Platinum Group Metals (PGMS): Palladium, Platinum, Rhodium, Iridium, Osmium, Ruthenium, are often in use for environmental technologies, consumer electronics, health, steel-making, defence, space exploration, and aviation. These materials are not only 'critical' for key industry sectors and future applications, but also for the sustainable functioning of the European economy. Hence, it is crucial to find a way in order to secure concentration of CRM. One of the directions postulated in this paper is to produce high temperature coatings using thermal spray technique like High Velocity Oxy Fuel (HVOF) for harsh application based on recycling technology that consumes post machined chips out of austenitic steels and Ni-based alloys with more than 15 wt% of Cr in metal matrix.

Currently one of the leading thermal spray processes to use in order to enhance corrosion resistance of low alloyed steels is HVOF process [11]. The technology produces coatings with high thickness (> 50 μm) with good bonding structure to the metallic core [12]. Nevertheless, the powder used to a coating production originates throughout expensive long term technological processes from natural resources (metal ores) leading to deplete of those resources extensively, hence development of powders for coatings is extremely costly and resource intensive. Moreover, each individual thermal spray process needs a high quantity of powder to deposit a 1 m^2 area. Sadly, thermal spray processes are not efficient as should be i.e. for deposition of WC/Co 200 – 300 μm thick coating on 1m^2 area at least 12 – 14 kg of powder is needed, to deposit 100 μm thick coating based on austenitic steel (FeCrNi powder) on 1 m^2 area at least 3 kg of powder is needed. It is important to note, that powder usage in thermal spray processes is dictated by individual factors such as density of powder, thickness of coating to be deposited, torch design, temperature, distance from torch to the substrate and more, hence it is difficult to measure exact amount of powder needed for a coating deposition. Hence, in this study an idea is to produce, deposit and test at high temperatures the coatings produced from metallic chips of 304L and Haynes® 282® Ni-based alloy. Oxidation tests in static air under atmospheric pressure were conducted at 500 °C for plain steel with 304L based coating and at 700 °C for Haynes® 282® coating type deposited on 304L steel surface. Prior and post examinations were carried out by standard analytical equipment using: single lens digital camera Canon 70D coupled with Canon MP-E 65 mm f/2,8 macro lens, Optical Microscope, Scanning Electron Microscope (SEM) Hitachi TM3000 equipped with Energy Dispersive X-Ray Spectrometry (EDS) Quantax 70 analyser from Bruker.

2. Experimental procedure

In this work two main coating systems were produced by HVOF technology and tested under air oxidation conditions: 304L and Haynes® 282® type coatings with the following composition.

Table 1 Chemical composition of 304L steel chips used for coating production

(wt%)	Fe	Cr	Ni	Mn	S	P	C	N	Si
304L steel	Bal.	19	10	2	0.003	0.045	0.03	0.1	0.75

Table 2 Chemical composition of Haynes® 282® steel chips used for coating production

(wt%)	Ni	Fe	Cr	Co	Mo	Si	Mn	Cu	La	Ti	Al	C	W	B
Haynes® 282®	Bal.	1.5	20	10	8.5	0.15	0.3	-	-	2.1	1.5	0.06	-	0.05

The fine grain powders were produced using liquid-assisted process for consolidation of metallic chips by melting and casting into the ingots. The processes are described in details under two application patents entitled: Development route in order to form fine grain powder using recycling technology of Fe-based chips and Development route in order to form fine grain powder using recycling technology of Ni-based chips. Patent application numbers: P. 417489, P. 417490. The presented technology roughly is presented below:

- Post processing chips delivery,
- Degreasing process using available ultrasonic bath at 40 °C for at least 40 minutes (x3 times), chemical composition analyses by X-ray Fluorescence (XRF) and Energy-dispersive X-ray spectroscopy (EDS) analyses),
- Liquid-assisted process for consolidation of metallic chips by melting and casting into the ingots,
- Casting ingots formation, chemical composition analyses by X-ray Fluorescence (XRF) and Energy-dispersive X-ray spectroscopy (EDS) analyses,
- Atomisation of the produced ingots into fine grain powder, chemical composition analyses by X-ray Fluorescence (XRF) and Energy-dispersive X-ray spectroscopy (EDS) analyses,
- Development of coating by HVOF process.

The process shown above is described in **Figure 1** for better understanding.

IOP Conf. Series: Materials Science and Engineering **329** (2018) 012010 doi:10.1088/1757-899X/329/1/012010

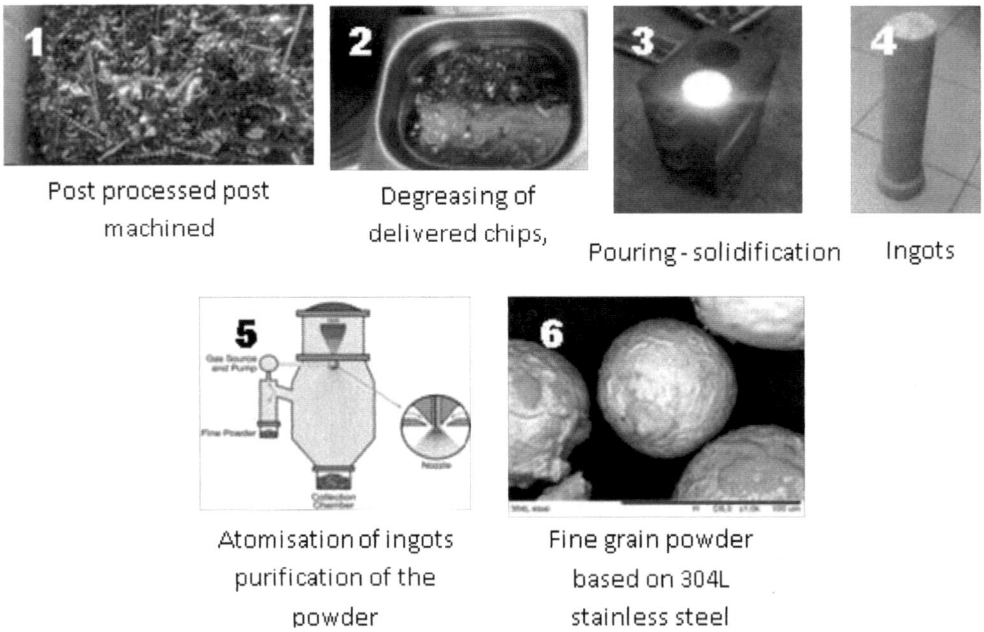

Figure 3 Development of fine grain powder from post processing chips using recycling technology

The produced powders were chemically analysed in order to investigate chemical composition changes after liquid-assisted process for consolidation of metallic chips by melting and casting into the ingots. The results for 304L and Haynes® 282® are presented in **Table 3** and **Table 4** respectively.

Table 3 Chemical composition performed using EDS analyser of the as deposited coating based on 304L chips powder

Spectrum: Point

Element	AN	Series	unn. C [wt.%]	norm. C [wt.%]	Atom. C [at.%]	Error [%]
Iron	26	K-series	27.29	39.25	25.23	0.8
Chromium	24	K-series	16.57	23.84	16.46	0.5
Oxygen	8	K-series	14.64	21.06	47.26	2.6
Nickel	28	K-series	9.62	13.83	8.46	0.3
Silicon	14	K-series	1.41	2.03	2.59	0.1
		Total:	69,53	100,00	100,00	

Table 4 Chemical composition performed using EDS analyser of the as deposited coating based on Haynes® 282® chips powder

Spectrum: Point

Element	AN	Series	unn. C [wt.%]	norm. C [wt.%]	Atom. C [at.%]	Error [%]
Nickel	28	K-series	30.01	35.10	21.51	0.9
Oxygen	8	K-series	18.67	21.83	49.08	3.1
Chromium	24	K-series	18.24	21.34	14.76	0.5
Molybdenum	42	L-series	5.92	6.93	2.60	0.2
Cobalt	27	K-series	5.84	6.83	4.17	0.2
Aluminium	13	K-series	2.96	3.47	4.62	0.2
Titanium	22	K-series	2.96	3.47	2.60	0.1
Iron	26	K-series	0.88	1.03	0.67	0.1
		Total:	85,49	100,00	100,00	

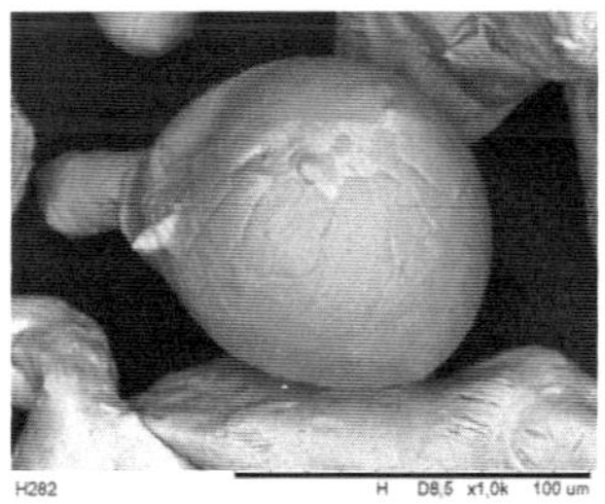

The results presented above clearly indicate, that both powders during fabrication underwent oxidation process, content of oxygen in both powders are quite high and reached up to 25 wt%. The powders showed high degree of oxidation during the liquid-assisted process for consolidation of metallic chips, perhaps due to easy access of oxygen from an ambient atmosphere. However, the problem can be avoided in the future when vacuum furnace with a protective atmosphere will be used. Nevertheless, it is important to mention that Cr content in both powders were high (over 20 wt%). High Cr content is crucial for development of a protective oxide scale in oxidising atmospheres, role of Cr in protective oxide formation is well known from many years [13, 14]. Finally fine grain powders that were produced via atomisation process under argon atmosphere [15] showed different dimensions ranging from 50 up to 300 μm. Both types of coatings were produced using High Velocity Oxy Fuel (HVOF) using certain facilities and under the following conditions:

- Gun: AMIL Werkstofftechnologie GmbH
- HVOF manipulator: GTV 2-AH1500/500

Prior the coatings deposition, powders were preheated by propane and oxygen to plastic or semi liquid/liquid state, further the powders were mixed with flammable gases and air. The HVOF process used compressed air pressure of 0.55 MPa, with distance from the surface of steel: 200 mm. Prior deposition, surface was cleaned in organic solutions and grinded to obtain roughness of R_{y5} = 45 μm.

3. Results and discussion

3.1 As deposited coatings

Figure 4 shows 304L coatings deposited by HVOF process on plain steel with different coating thickness, whereas **Figure 5** presents Haynes® 282® coatings that were deposited based on different fine grain powder dimensions deposited on 304L steel. The pictures were captured using single lens digital camera Canon 70D coupled with Canon MP-E 65 mm f/2,8 macro lens.

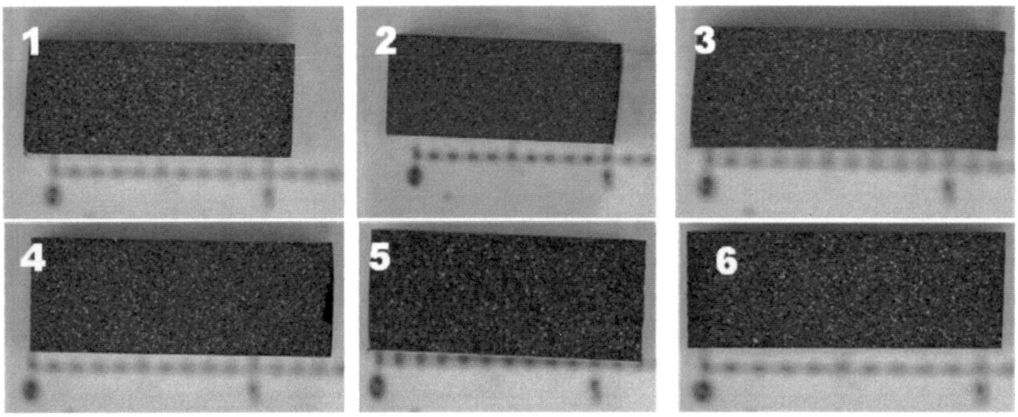

Figure 4 As deposited coatings based on 304L steel chips by HVOF process on plain steel (with different thickness)

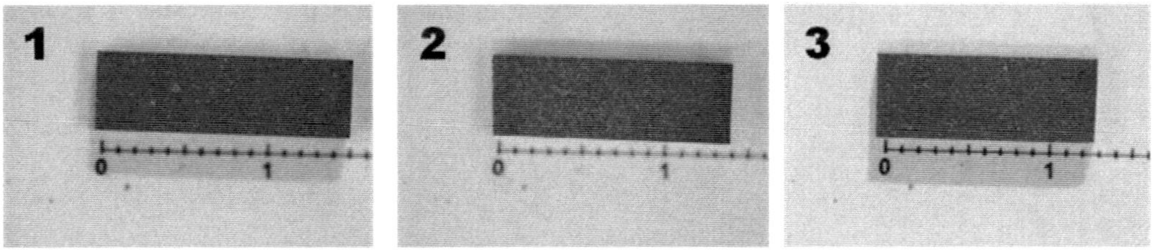

Figure 5 As deposited coatings based on Haynes® 282® steel chips by HVOF process on 304L steel: A) 45 μm, B) 45 – 100 μm, and C) 100 – 200 μm

The coatings showed in **Figure 4** and **Figure 5** respectively, present, no cracks of the deposited layers, suggesting good bond strength that formed between the substrate and the deposited coatings when HVOF process was used. Moreover, the analyses performed by EDS on as deposited coatings that were carried out shown no much difference between the chemical compositions of the powders that were used, suggesting that HVOF process showed no influence on chemical composition of the developed coatings.

3.2 Kinetic behaviour

The samples with dimension of approximately 1.5, 0.7 and 0.4 cm were decreased using organic solution and ultrasonic bath for 15 minutes at 40 °C and accurately weight using digital balance CPA 225D-0CE prior air oxidation test and during exposures. As mentioned previously air oxidation test for 304L coatings on plain steel was carried out for 100 hours at 500 °C, but coatings based on Haynes® 282® post machined chips deposited on 304L were exposed at 700 °C for 300 hours. Together with the coated materials, the uncoated plain steel and 304L steel was exposed as a reference sample. The kinetic data shown in **Figure 6** and **Figure 7** from high temperature tests showed that the coatings made of post machined scrap indicate some rate of protectiveness.

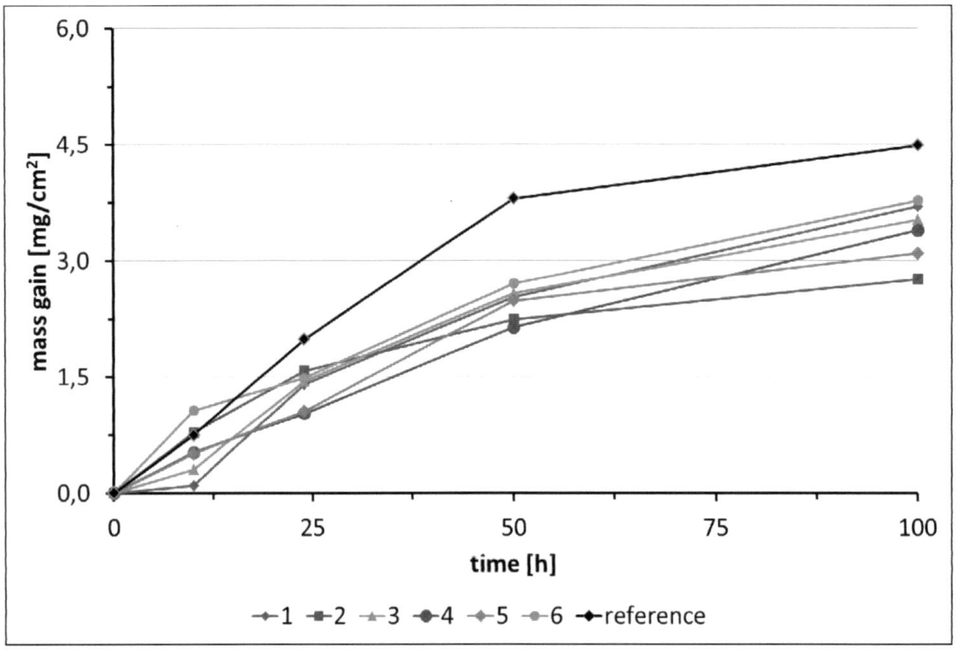

Figure 6 Kinetic data of the exposed 304L coating based on metallic chips deposited on plain steel at 500 °C for 100 hours

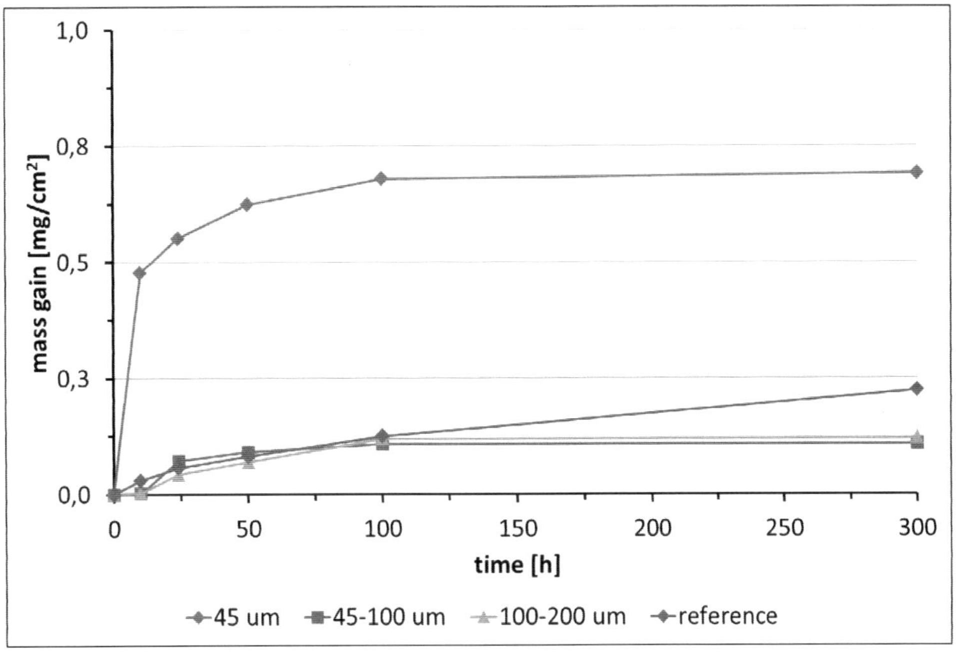

Figure 7 Kinetic data of the exposed Haynes® 282® coating based on metallic chips deposited on 304L steel at 700 °C for 300 hours

Kinetic data presented in **Figure 6** shows, that plain steel coated by 304L based coating indicate some degree of protectiveness. The lowest mass gain was obtained by sample 2, the highest mass gain was found to be in the reference sample (plain steel). Its need to be pointed out that in the initial stages of oxidation, *sample 6*, showed higher mass gain than that observed in reference sample, but after longer exposure mass gain of the reference sample reached the highest values (4.5 mg/cm^2). Enhanced corrosion resistance that was observed in this work for coated plain steel is related mainly to the higher Cr content in the deposited coating, that induces development of the oxide scale that possess higher corrosion resistance against oxidation atmosphere. Furthermore, the coated samples showed similar mass gain after 100 hours at 500 °C exposure that suggests that thickness of the deposited coatings show no much influence on the oxidation kinetics. In contrast to the data showed in **Figure 6**, in **Figure 7** kinetic data of the exposed Haynes® 282® coating developed on the basis of post machined chips and deposited on 304L steel surface indicates that the coating with 45 µm grain dimension showed the highest mass gain. In contrast, two other coatings (45 - 100 µm and 100 – 200 µm grain size), indicated lower mass gain than that presented by reference material (the uncoated 304L steel). Finally, the kinetic curve that developed at 700 °C for 300 hours for the uncoated sample indicate constant weigh growth in contrast to two coatings (45 - 100 µm and 100 – 200 µm) where steady state oxidation process occurred.

3.3 Surface observations

Figure 8 shows macro images captured after exposure at 500 °C for 100 hours for plain steel coated with 304L based steel.

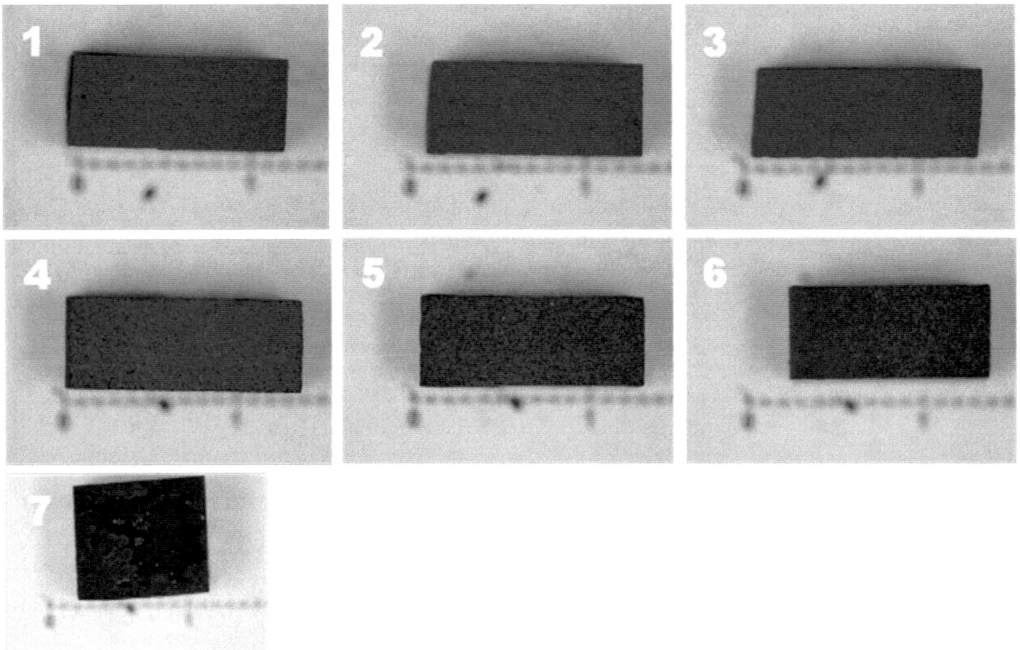

Figure 8 Macro images captured after exposure at 500 °C for 100 hours for plain steel coated with 304L based steel (samples 1 – 6 coating with different thickness), 7) the uncoated plain steel

Macro images shown in **Figure 8** clearly indicate and confirms kinetic data showed in **Figure 6**, the higher mass gain of the samples is closely related to iron oxide formation, perhaps due to thin or inadequate coating process. As illustrated, *samples 1 – 3* show no hematite formation (red colour), *sample 4* is slightly affected by hematite formation, whereas *sample 5* and *sample 6* showed similar surface occurrence as the reference sample (plain steel), that is rich in red hematite phase on the surface [16]. It is well known that iron based oxides are less protective at high temperatures than the Cr based oxides where diffusion processes are much quicker due to high number of defects inducing higher kinetic growth [17]. **Figure 9** shows macro images of the exposed Haynes® 282® coating based on metallic chips deposited on 304L steel exposed at 700 °C for 300 hours.

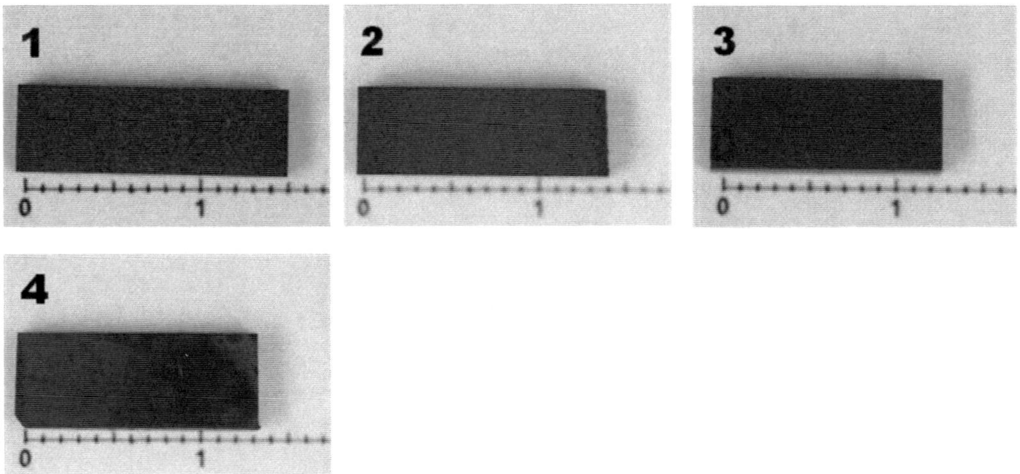

Figure 9 Macro images captured after exposure at 700 °C for 300 hours for Haynes® 282® coated 304L steel: 1) 45 μm size powder, 2) 45-100 μm powder size, C) 100-200 μm powder size, and D) the uncoated 304L steel

In contrast to observations for the coated plain steel, in current situation deposition of Haynes® 282® coating on 304L steel induced reduction in hematite development due to higher Cr content in the coating and in the substrate material. Under oxidising atmosphere rich Cr coating (Haynes® 282®) and substrate (304L) provoked higher rate of Cr outward diffusion, responsible for developing of highly resistant oxide scale that consisted: fraction of Cr_2O_3 oxide, $FeCr_2O_4$, Ni-Fe-Cr_2O_4 spinel. The formation of the mentioned phases in austenitic steels are broadly described by many authors in the past, therefore a reader is directed to the following papers [14, 18, 19, 20, 21].

3.4 Cross-section observations

Figure 10 shows cross-sectioned, optical microscope images of the exposed 304L coatings deposited on plain steel at 500 °C for 100 hours. As is, the samples 1 – 3 are covered by thick 300 μm coating based on 304L composition, the other 3 samples (4 – 6), are coated by a thin coatings, whereas 7 showed the uncoated plain steel. The coated materials showed no cracks formation, no detachment or delamination of the deposited coating. The results in this study suggest that even deposition of a thin coating is beneficiary for high temperature exposure in air conditions due to the formation of more protective oxide scale than that developed on the uncoated material. The mechanism of surface protection relays on Cr diffusion from the coating that contains 20 wt% of Cr to the external part of the coating and development of Cr_2O_3 rich oxide scale. The oxide in rich oxygen atmosphere (air) due to more negative value of Gibbs free energy formation for Cr_2O_3 than that for Fe-based oxide, is more thermodynamically stable and offers higher corrosion resistance than that obtainable by Fe-based oxide.

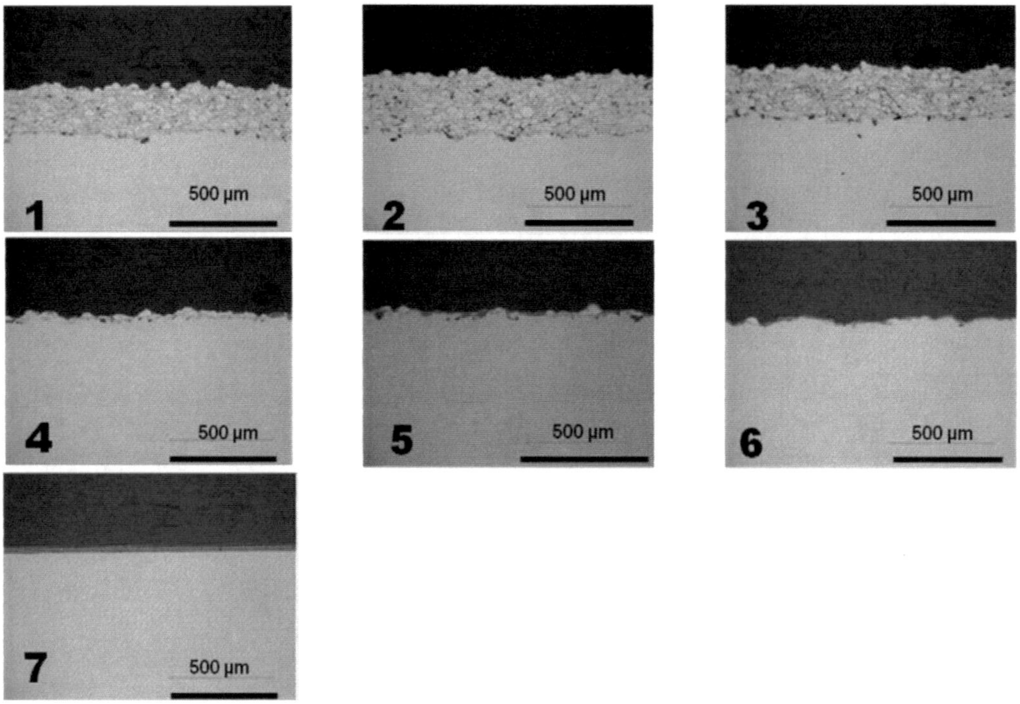

Figure 10 Cross-sectioned microstructures of the exposed 304L coating on plain steel exposed at 500 °C for 100 hours

In case of Haynes® 282® coating, surprisingly, the thickest coating developed using 45 µm fine grain size showed the poorest corrosion resistance in terms of oxidation protection. The thick coating (>500 µm) that was applied on 304L steel according to EDS analyses, showed high iron content suggesting high diffusion rate of iron from the 304L substrate throughout the coating. The other two coatings on cross-section images presented in **Figure 10** (sample 2 and samples 3) indicate that during deposition by HVOF, large powder (grain size) showed no adherence to the 304L steel surface suggesting that kinetic energy of the powder during deposition process was to high resulting that powder was ejected instead of adhered to the 304L steel substrate. The described process indicate effect of shot peening process, that is a cold working process that is often used in order to obtain a compressive residual stress layer and modify mechanical properties of alloys and composites. In addition some researchers proposed that shot peening improve corrosion resistance throughout the formation of protective layer on the surface of the material [22, 23] by diffusion of Cr ions from the metal matrix to the exposed surface, resulting development of Cr-rich region on the surface, that facilitate the formation of dense, protective, well adherent Cr_2O_3, film [24].

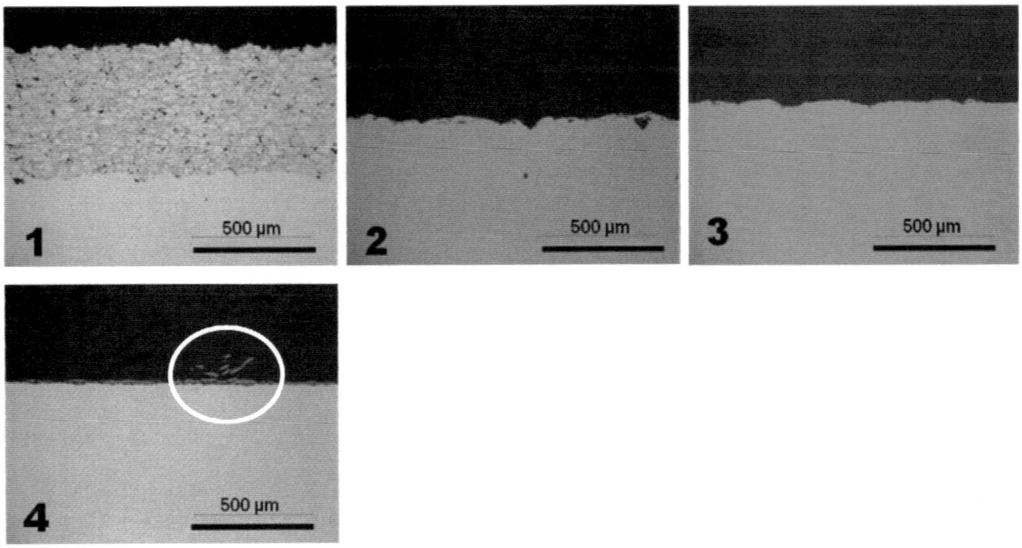

Figure 11 Cross-sectioned microstructures of the exposed Haynes® 282® coating on 304L steel exposed at 700 °C for 300 hours

The uncoated 304L steel developed under air oxidation process at 700 °C for 300 hours, a thin oxide scale that is composed mainly from $FeCr_2O_4$ spinel phase with some areas enriched in Cr suggesting the formation of Cr_2O_3 oxide. Nevertheless, the steel under relatively short exposure indicate some corrosion degradation due to brittle, fractured scale that formed (white circle). In overall, the corrosion resistance of the uncoated 304L steel can be compared to the materials that were coated by the coating with the thickest grains. Moreover, it can be suggested, based on the achieved results and cross-section images, that in the case of thick grain, effect of shot peening is stronger than influence of Haynes® 282® coating that was applied based on post machined scrap.

4. Conclusions

The aim of this paper was to show a solution for Critical Raw Materials (CRM) such as Cr by the development of fine grain powders derived from post machined chips rich in Cr.

Chromium (Cr) as a one of Critical Raw Material is often used as an alloying element to steels, Ni-based alloys and coatings for better corrosion protection; hence, concentration of Cr in earth is significantly reduced. Nevertheless, Cr can be regained to secure CRM by the simple process.

The chips may originate either from austenitic steels or Ni-based alloys. In general post machined chips are treated frequently as a scrap material with no value, especially when are contaminated by organic based lubricant. However, in this work post machined chips after liquid-assisted process for consolidation of metallic chips by melting and casting into the ingots, atomisation, and thermal spray HVOF process were deposited on two different materials.

The produced powders based on 304L stainless steel and Haynes® 282® Ni-based alloy post machined chips were deposited on a plain steel and 304L steel respectively. Both type of corrosion couples, where exposed in air oxidation tests for 100 and 300 hours at 500 and 700 °C respectively. The research carried out in this study showed, that coatings produced using thermal spray process exposed at high temperatures offer protection for plain steel, as well as for 304L steel. Nevertheless, it is need to pointed out, that powder size is crucial for coating deposition process. It was found that due to large grain size of powder, instead of deposition, ejection of powder appeared, leading to the formation of shot peeing effect. Nevertheless, the presented results in this study, can be an excellent starting point for further work that can be carried out in the field of high temperature protection where post machined wastes can be re-used in order to secure Critical Raw Materials (CRM) like Cr. Furthermore, the produced powder can be used as well for Hot Isostatic Pressing (HIP) process and in 3D printing technology such as: SLM – Selective laser melting.

Acknowledgement

Authors like to acknowledge the support of fundamental research performed in this study to Statutory Project funded from Foundry Research Institute, project number: 7010/00. The author of the work acknowledges as well post machined chips suppliers: Haynes International USA and Ceramit Poland

References

[1] Economic analysis of boiler tube failures, Economic Analysis of Boiler Tube Failures in NSW, RF Small, Connell Wagner PPI, 116 Military Rd, Neutral Bay, NSW 2089 Australia
[2] Dudziak T, Hussain T Simms N J Syed A U Oakey J E 2014 *Corros. Sci.,* **79** 184
[3] Dudziak T Grobauer S Simms N J Krupp U Lukaszewicz M Oakey J. E 2015 *High Temp. Mater. Processes,* **34(8)** 783
[4] Viswanathan R Foulds J R Roberts D I Pro*ceedings on Boiler Tube Failures in Fossil Power Plants,*1987 **10** 3 35
[5] Viswanathan R Foulds J R **1995** *ASME Pressure Vessels and Piping Conference* **303** 187
[6] Viswanathan R Damage Mechanisms and Life Assessment of High Temperature Components. Metals Park: ASM International; 1989
[7] Lamping G.A Arrowwood Jr. R.M 1985 Manual for Investigation and Correction of Boiler Tube Failures. Report CS-3945. Palo Alto: Electric Power Research Institute (EPRI)
[8] Fürbeth W, Schütze M, 2009 *Mater. Corros.* **60**, 481
[9] General budget of the European union for the financial year 2009, European Commission, Luxemburg 2009
[10] Power Plant Life Management and Performance Improvement, Woodhead Publishing, 2011 - Technology & Engineering, edited. J. E. Oakey
[11] High Velocity Oxy-Fuel (HVOF) Solutions, Sulzer Metco Booklet, Issue 5, (2013)
[12] Tucker, Jr R. C Thermal Spray Coatings, ASM Handbook, Volume 5: Surface Engineering Cotell C M Sprague J A Smidt, Jr F A. editors, 497
[13] Henry J Zhou G Ward T 2007 *Mater. High Temp.,* **24** 249
[14] Lepingle V Louis G Petelot D Lefebvre B Vaillant J.C 2001, *Mater. Sci. Forum.,* **369-372**, 239
[15] Lagutkin S Achelisa, L Sheikhaliev S Uhlenwinkel V Srivastava V 2004 *Mater. Sci. Eng., A* **383** **1** 1
[16] http://geology.com/minerals/hematite.shtml
[17] Limoge Y Bocquet J. L 2001 *Defect Diffus Forum* **194–199** 1051
[18] Wright I. G Pint B. A An assessment of the High-Temperature Oxidation Behaviour of Fe-Cr Steels In Water Vapour And Steam, Corrosion 2002 Conference Denver, USA
[19] Chen R.Y Yeun W.Y.D 2003 *Oxid. Met.,* **59(5-6)** 433
[20] Ehlers J. Young D.J. Smaardijk E.J Tyagi A.K, Penkalla H.J, Singheiser L Quadakkers W.J 2006 *Corros. Sci.,* **48** 3428
[21] Sundararajan T Kuroda S Kawakita J Seal S 2006 *Surf. Coat. Technol.* **201** 2124
[22] Naraparaju R Trindade V B Christ H.-J Krupp U 2009 *Corros Eng Sci Techn.,* **44.3** 211
[23] Xingeng L Jiawen H 2006 *Mater. Lett.,* **60.3** 339
[24] Minami Y 1996 *NKK Tech Rev.,* **75** 1

IOP Conf. Series: Materials Science and Engineering **329** (2018) 012011 doi:10.1088/1757-899X/329/1/012011

Structural transformations of heat treated Co-less high entropy alloys

D Mitrica [1*], A Tudor[1], A Rinaldi[2], V Soare[1], C Predescu[3], A Berbecaru[3] and F Stoiciu[1] , V Badilita[1]

[1]National R&D Institute for Nonferrous and Rare Metals – IMNR, 102 Biruintei Blvd, Pantelimon, Ilfov, Romania.

[2]Agenzia Nazionale per le Nuove Technologie, l'Energia e lo Sviluppo Economico Sostenabile-ENEA, Casaccia Research Centre, Italy

[3]University POLITEHNICA of Bucharest, 313 Splaiul Independenței, 6 District, 060042 Code, Bucharest, Romania.

dmitrica@imnr.ro

Abstract. Co is considered to be one of the main ingredients in superalloys. Co is considered a critical element and its substitution is difficult due to its unique ability to form high temperature stable structures with high mechanical and corrosion/oxidation resistance. High entropy alloys (HEA) represent a relatively new concept in material design. HEA are characterised by a high number of alloying elements, in unusually high proportion. Due to their specific particularities, high entropy alloys tend to form predominant solid solution structures that develop potentially high chemical, physical and mechanical properties. Present paper is studying Co-less high entropy alloys with high potential in severe environment applications. The high entropy alloys based on Al-Cr-Fe-Mn-Ni system were prepared by induction melting and casting under protective atmosphere. The as-cast specimens were heat treated at various temperatures to determine the structure and property behaviour. Samples taken before and after heat treatment were investigated for chemical, physical, structural and mechanical characteristics. Sigma phase composition and heat treatment parameters had major influence over the resulted alloy structure and properties.

1. Introduction

High entropy alloys (HEA) are multicomponent alloys composed of five or more elements in equal or near equal proportions. Due to the Boltzmann's hypothesis, where the maximum entropy of mixing of an alloy is obtained at equiatomic compositions, HEAs are expected to form only simple solid solution structures and develop superior mechanical and corrosion properties [1].

There is a substantial amount of papers that describe the thermodynamic intricacies related to solid solution formation in high entropy alloys. Significant criteria were developed based on the entropy, enthalpy, atomic size difference, valence electron configuration and electronegativity of various high entropy alloy systems [2-4], which are contributing to the selection of viable alloy compositions for different applications. Computer-Aided Phase Prediction in high entropy alloys was also performed on various HEA systems, using specialized methods such as Calphad, Dendity Functional Theory, Monte Carlo or Molecular Dynamics, with significant results in approximating phase constitution and evolution at different compositions and temperatures [5-10],.

Experimental trials were performed by various authors for a large number of alloy compositions containing mostly transitional elements (Co, Cr, Fe, Cu, Ni, Ti, Zr,etc.) with controlled additions of Al, Si, B or C for property enhancement[11-13]. Among all the Al-Co-Cr-Fe-Ni is the most studied system [14–16]. Co as a Face Centered Cubic (FCC) forming element was shown to produce stable solid solution structures with high mechanical and corrosion properties. Co was also seen to enhance mechanical resistance and hardness of HEA after annealing at up to 600°C. Nevertheless Co is a well-known critical element and its replacement is desirable in future industrial materials.

. High entropy alloys based on Al-Cr-Fe-Mn-Ni system was shown to have interesting properties, especially after heat treatment at 800°C [17]. Aluminum was used to tailor the FCC to BCC transition in the newly developed alloy. Hardness values of up to 900Hv were obtained for the heat treated alloys.

Present paper is studying several high entropy alloys without Co content, based on Al-Cr-Fe-Mn-Ni system, taking into account variable solidification rates and heat treatment temperatures. A correlation of criteria, experimental and modelling results was performed on selected alloys.

2. Experimental

High entropy alloys based on Al-Cr-Fe-Mn-Ni system were prepared in an induction furnace type Linn MFG – 30 with inert atmosphere. Technical purity (99.9%) elemental Al, Cr, Fe, Mn and Ni were used as raw materials. A 500g charge of each alloy composition was initially loaded and melted under argon atmosphere, in a zirconia based crucible. The alloy was re-melted to ensure a uniform chemical composition and was cast at different solidification rates in graphite, steel and copper molds. The as-cast alloys were annealed in an electrical furnace, LHT 04/17 Nabertherm GMBH under protective atmosphere (Ar). The heat treatment stage was conducted at temperatures between 500 and 1100°C for 50 hours, with slow furnace cooling. Samples were taken before and after the heat treatment process for chemical, structural and mechanical analyses.

The chemical composition of the alloy was determined by optical emission spectrometry in inductively coupled (ICP-OES) using an Agilent 725 spectrometer. Optical microscopy investigation was performed with a Zeiss Axio Scope A1m Imager microscope. Samples were previously etched in an HCl-HNO3-CH3COOH-H2O solution to enhance the visibility of the grains and the grain boundaries. The morphology of the alloy was analysed by scanning electron microscopy (SEM) using a FEI Quanta 3D FEG operating at 20-30 kV, equipped with an energy dispersive X-ray spectrometer (EDS). The phase structure was analysed by X-ray diffractometry (XRD). Data acquisition was performed on BRUKER D8 ADVANCE diffractometer, using Bruker®DIFFRAC plus software, Bragg-Brentano diffraction method, Θ - Θ coupled in vertical configuration, with the following parameters: CuKα radiation, 2Θ Region: 20 ÷ 1240, 2Θ Step: 0.020, Time/step: 8.7 sec/step. Vickers microhardness of the samples was measured at room temperature using an optical microscope attachment (Anton Paar MHT10), with an applied load of 2N and slope of 0.6 N/s. In order to determine the average hardness of the alloy, seven indentations were performed on the surface of the as-cast and heat treated samples.

3. Results and discussion

3.1. Criteria calculations

The criteria for phase formation in high entropy alloys have been intensely studied in the past few years by various researchers [2-4,13,18-26]. Based on the definition of HEA concept, the alloys with mixing entropy (ΔS_{mix}) higher than the intermetallic compound formation entropy(ΔS_{IM}) can form mainly solid solution structures. Later, the mixing entropy was found to be defined mostly by the configurational entropy (ΔS_{conf}) [13], that needed to be higher than 11 J/moleK and lower than 19.5

J/moleK [18, 19]. Taking into account the definition of Gibbs free energy, in order to arrive to a stable alloy configuration is necessary to have a balanced contribution between configuration entropy and mixing enthalpy of the system (ΔH_{mix}). Solid solution formation was also studied based on Hume-Rothery rule, where the atomic-size difference (δ), electronegativity($\Delta\chi$) and valence electron concentration (*VEC*) play an important role in alloy selection [20,21]. A $\Delta H_{mix} - \delta$ criterion was established [18, 19], where an enthalpy of mixing between -11.6 kJ/mol and 3.2 kJ/mol, and an atomic size difference smaller than 6.6% would define a system containing mainly solid solution structures. In order to take in consideration the entropy effect, an Ω factor was defined as the ratio between ΔS_{mix} and ΔH_{mix}, and is considered only together with δ. Thus if Ω >1.1 and δ < 3.6% only solid solutions are formed, if 1.1< Ω < 10 and 3.6% < δ < 6.6% solid solutions and intermetallic compounds are formed, and if Ω > 10 only solid solutions are formed [22]. The electronegativity (Allen) effect was also considered and values between 3% and 6% showed potential for the formation of solid solutions in multicomponent alloys [23]. The type of solid solutions formed in high entropy alloys was determined based on the valence electron concentration (*VEC*), where values smaller than 6.87 indicated the formation of only Body Centered Cubic (BCC) solid solutions, values between 6.87 and 8 showed that Body Centered Cubic (BCC) and Face Centered Cubic (FCC) structures coexist, and if VEC > 8 only Face Centered Cubic (FCC) structures were formed [24].

Due to the contradicting results obtained through the experimental process the criteria was further developed by other authors taking into consideration the effect of intermetallic phases formation more than solid solution stability criteria. Thus it was determined that a content larger than 40 at.% of paired sigma-forming element (PSFE) is decisive on the formation of sigma phase in high entropy alloys [25,26]. If 25 at. %< PSFE < 40 at.% sigma phase may not exist and if PSFE < 20 at% sigma phase should not be present in the alloy structure. The influence of intermetallic phases was also studied by other authors [3,4] that determined a k_1^{cr} correlation factor ($k_1^{cr} \equiv 1 - \frac{T_A \Delta S_{mix}}{\Delta H_{mix}}(1 - k_2) > \frac{\Delta H_{IM}}{\Delta H_{mix}}$) that needs to be higher than the ratio between intermetallic and solid solutions mixing enthalpies, in order to form only solid solutions.

The calculation of criteria values for some alloy compositions derived from Al-Cr-Fe-Mn-Ni system are presented in table 1.

Table 1. Criteria calculations for HEA structure prediction

Alloy	**ΔS_{conf} J/molK**	**ΔH_{mix} kJ/mol**	**δ %**	**VEC %**	**$\Delta\chi$ %**	**Ω**	**PSFE**	**ΔH_{IM} kJ/mol**	**k_1^{cr}**	**$\Delta H_{IM}/\Delta H_{mix}$**
AlCrFeMnNi	13.38	-12.48	5.82	7	5.6	1.74	40	-27.84	1.5	2.23
$Al_{0.5}CrFeMnNi_{0.5}$	12.96	-8.25	5.28	7	4.95	2.66	50	-19	1.73	2.3
$Al_{0.5}Cr_{1.5}Fe_{1.5}MnNi_{0.5}$	12.51	-6.52	5.00	7	4.85	3.36	60	-15.52	1.9	2.38
$Al_{0.5}Cr_2Fe_{1.5}MnNi$	12.61	-6.97	4.72	7.16	5.21	3.22	50	-15.4	1.85	2.2
$Al_{0.3}CrFe_{1.5}MnNi_{0.5}$	12.32	-5.5	4.70	7.26	4.6	3.9	47	-13.42	2.05	2.43
$Al_{0.3}Cr_{1.5}Fe_{1.5}MnNi_{0.5}$	12.16	-4.94	4.53	7.12	4.71	4.39	63	-12.25	2.15	2.48

The results showed that most of the alloy compositions are characterized by the high entropy effect, with ΔS_{conf} and ΔH_{mix} close to the minimum interval limits, atomic size difference and electronegativity in required range and Ω showing a predominant formation of solid solution phases. BCC and FCC type of solid solutions are expected to form. The intermetallic formation criteria is confirming the previous findings with PSFE greater than 45 and $\Delta H_{IM}/\Delta H_{mix} > k_{cr}$ for most of the alloys. None of the alloy compositions are predicted to form only solid solution structures.

3.2. Experimental results

Three selected compositions, from Al-Cr-Fe-Mn-Ni system, are presented here in as-cast and heat treated state. The alloys presented a homogenous composition over the entire body of the resulted ingots (table 2).

IOP Conf. Series: Materials Science and Engineering **329** (2018) 012011 doi:10.1088/1757-899X/329/1/012011

Table 2. The alloys chemical composition.

Alloy	Composition, wt. %				
	Al	Cr	Fe	Mn	Ni
$Al_{0.5}Cr_{1.5}Fe_{1.5}MnNi_{0.5}$	5.41	24.9	33.4	23.2	12.8
$Al_{0.5}Cr_{2}Fe_{1.5}MnNi$	3.82	30.6	27.3	18.2	19.8
$Al_{0.3}Cr_{1.5}Fe_{1.5}MnNi_{0.5}$	3.15	30.7	33.6	20.3	12.2

Optical microscopy results (figure 1) for the as-cast alloys showed a predominant dendritic structure with larger amount of hard compounds in the $Al_{0.5}Cr_{1.5}Fe_{1.5}MnNi_{0.5}$ alloy, local interdendritic elemental segregations for the $Al_{0.5}Cr_{2}Fe_{1.5}MnNi$ alloy and the presence of distinct rounded phases in the $Al_{0.3}Cr_{1.5}Fe_{1.5}MnNi_{0.5}$ alloy. Higher magnification, by SEM analyses (figure 2), revealed small and evenly distributed precipitated phases all across the alloys structure. The precipitates are larger and have different shapes in the $Al_{0.5}Cr_{2}Fe_{1.5}MnNi$ alloy. XRD analyses (figure 3) revealed only BCC phases in $Al_{0.5}Cr_{1.5}Fe_{1.5}MnNi_{0.5}$ alloy (A2 and B2 types). The $Al_{0.5}Cr_{2}Fe_{1.5}MnNi$ alloy showed a more complex structure with predominant BCC phases (A2 and B2 types) and minor FCC phase (A1 type). A significant proportion of sigma phase (20%) together with a BCC-A2 type phase was detected in the $Al_{0.3}Cr_{1.5}Fe_{1.5}MnNi_{0.5}$ alloy.

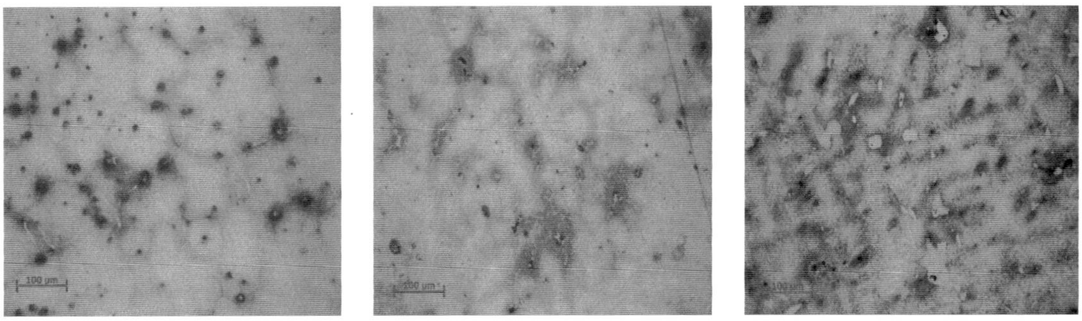

Figure 1. Optical micrographs of as-cast $Al_{0.5}Cr_{1.5}Fe_{1.5}MnNi_{0.5}$, $Al_{0.5}Cr_{2}Fe_{1.5}MnNi$ and $Al_{0.3}Cr_{1.5}Fe_{1.5}MnNi_{0.5}$ alloys.

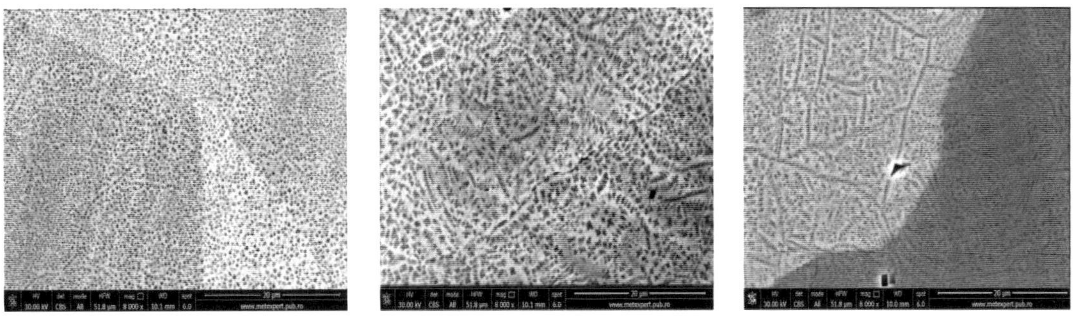

Figure 2. SEM image of as-cast $Al_{0.5}Cr_{1.5}Fe_{1.5}MnNi_{0.5}$, $Al_{0.5}Cr_{2}Fe_{1.5}MnNi$ and $Al_{0.3}Cr_{1.5}Fe_{1.5}MnNi_{0.5}$ alloys.

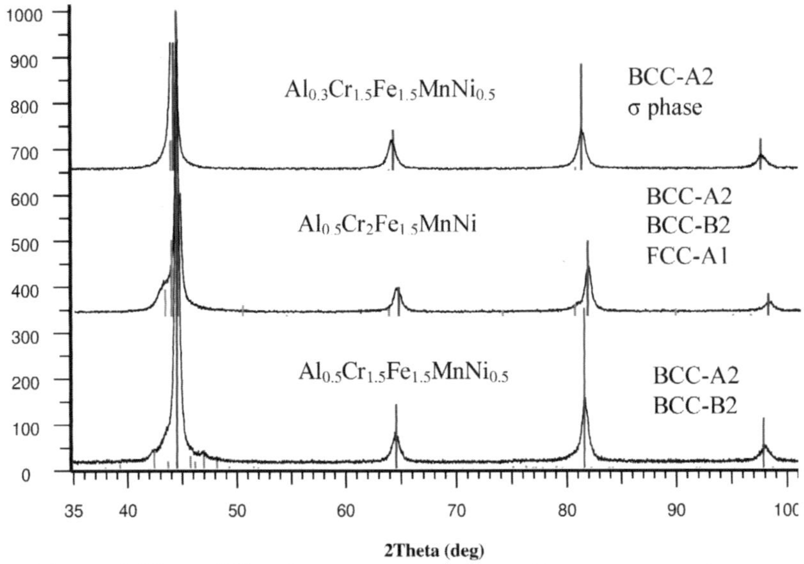

Figure 3. X-ray diffraction pattern and phases for the as-cast alloys

The structural behavior of $Al_{0.3}Cr_{1.5}Fe_{1.5}MnNi_{0.5}$ alloy was studied at different solidification rates. Optical analyses (figure 4) of the structure revealed a smaller amount of hard phases in the interdendritic area and larger light colour interdendritic phases with the increase in the solidification rate. No significant changes were shown for the dendrite configuration.

.

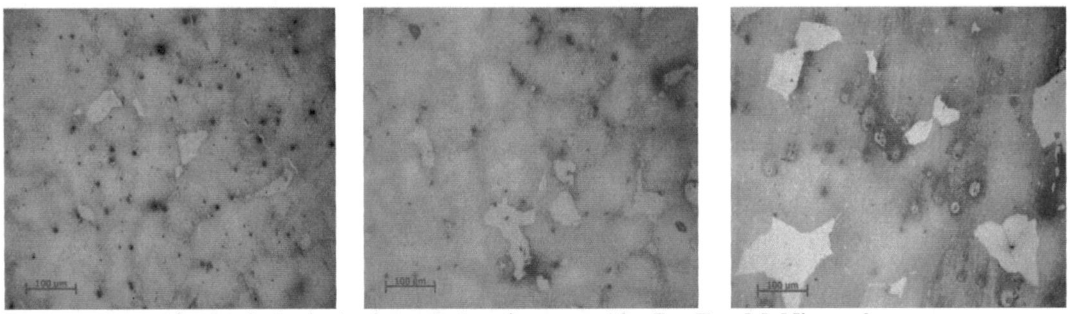

Figure 4. Optical micrographs of as-cast $Al_{0.3}Cr_{1.5}Fe_{1.5}MnNi_{0.5}$ at low, medium and high solidification rates.

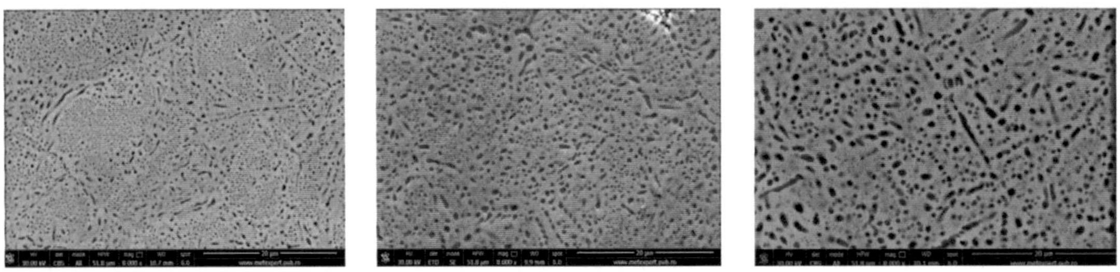

Figure 5. SEM images of heat treated $Al_{0.3}Cr_{1.5}Fe_{1.5}MnNi_{0.5}$ at 700, 800 and 900 °C

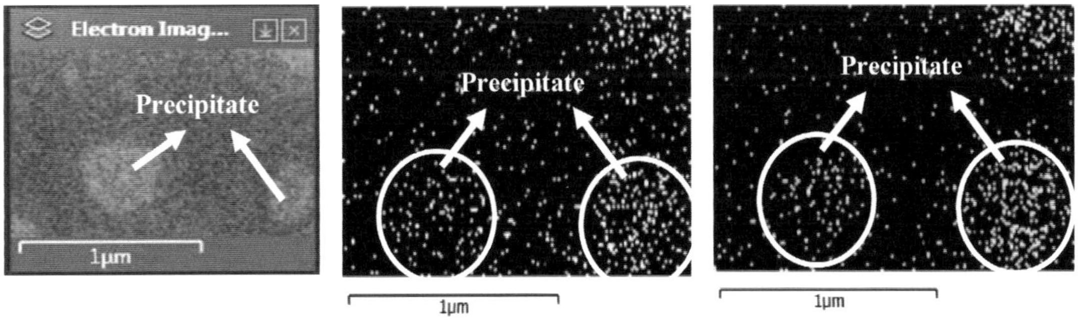

Figure 6. EDS mapping of $Al_{0.3}Cr_{1.5}Fe_{1.5}MnNi_{0.5}$ heat treated at 700°C

In figure 5 are presented SEM images of the heat treated alloys. The heat treatment process is shown to allow a substantial growth of the precipitates with the annealing temperature. The EDS mapping results (figure 6) shows a significant Al and Ni density in the precipitates region that suggests the formation of NiAl-B2 type phase. X-ray diffraction analyses (figures 7-9) revealed important phase changes between different annealing states of the alloys. An BCC-B2 and sigma phase structure was seen for the $Al_{0.5}Cr_{1.5}Fe_{1.5}MnNi_{0.5}$, obtained by the annealing at 700°C or 800°C. By annealing at 900°C, the BCC-A2 phase became more stable. The $Al_{0.5}Cr_{2}Fe_{1.5}MnNi$ alloy showed a structure formed of sigma phase and BCC-B2 at 700°C. At 800°C a FCC-A1 phase becomes stable. At 900°C a structure formed solely of BCC-A2 and FCC-A1 phases is revealed.

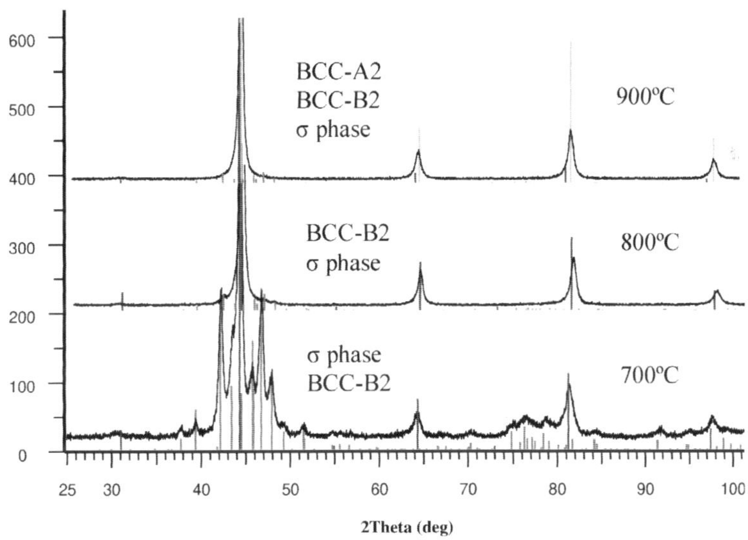

Figure 7. X-ray diffraction for $Al_{0.5}Cr_{1.5}Fe_{1.5}MnNi_{0.5}$ heat treated at 700, 800 and 900 °C

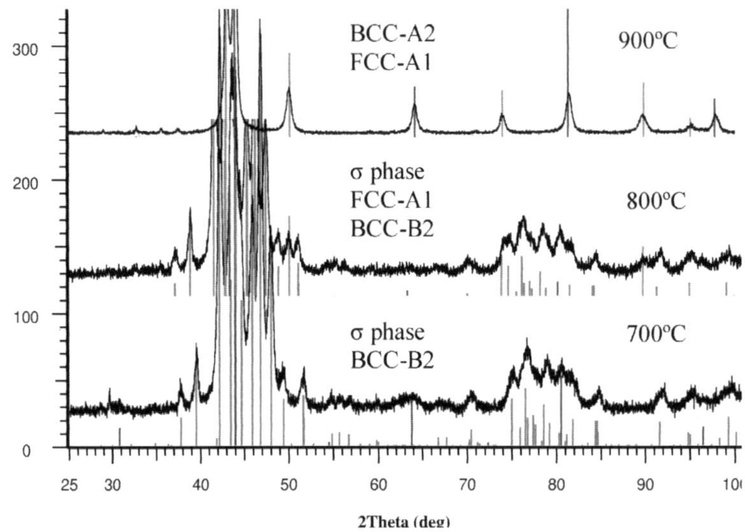

Figure 8. X-ray diffraction for $Al_{0.5}Cr_2Fe_{1.5}MnNi$ heat treated at 700, 800 and 900 °C

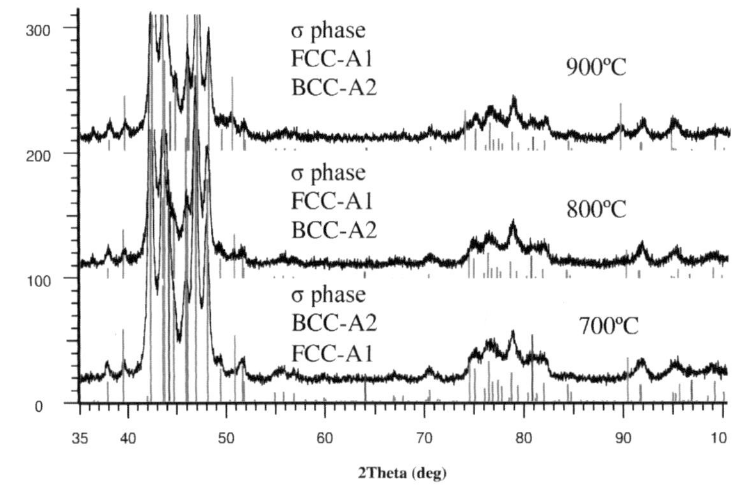

Figure 9. X-ray diffraction for $Al_{0.3}Cr_{1.5}Fe_{1.5}MnNi_{0.5}$ heat treated at 700, 800 and 900 °C

The $Al_{0.3}Cr_{1.5}Fe_{1.5}MnNi_{0.5}$ showed a similar phase composition between the three annealing temperatures. The sigma phase, BCC-A2 and FCC-A1 phases were detected in all the annealed structures. As a conclusion, one can determine that BCC-A2 phase is more stable at higher temperatures, that BCC-B2 phase is mainly stable at lower temperatures and that the alloys with lower aluminium content are prone to form FCC type structures. From EDS and XRD analysis one can see that sigma phase is the matrix and NiAl is the precipitated phase.

The hardness tests results (figure 10) showed a large increase for the $Al_{0.3}Cr_{1.5}Fe_{1.5}MnNi_{0.5}$ alloy, after heat treatment at 700, 800 and 900°C. The alloy hardness decreases strongly after an 1100°C heat treatment. Similar behaviour was indicated for the $Al_{0.5}Cr_2Fe_{1.5}MnNi$ alloy, nevertheless hardness decreases at as-cast levels after the heat treatment at 900 ° C. The $Al_{0.5}Cr_{1.5}Fe_{1.5}MnNi_{0.5}$ is showing no significant changes in hardness after the heat treatment process. By comparing the microhardness values with the phase constitution one can observe some abnormal behavior of $Al_{0.3}Cr_{1.5}Fe_{1.5}MnNi_{0.5}$ that even if contains a good proportion of FCC-A1 phase has the highest value among all three alloys. In this case the hard sigma phase has the highest proportion and might as well influence strongly the alloy mechanical properties. Same can be said about the $Al_{0.5}Cr_2Fe_{1.5}MnNi$ alloy that provides a much higher hardness in the annealed structures with high sigma phase content. Nevertheless, sigma phase presence in the $Al_{0.5}Cr_{1.5}Fe_{1.5}MnNi_{0.5}$ annealed at 700°C does not influence the microhardness value.

Microhardness measured for different solidification rates of $Al_{0.3}Cr_{1.5}Fe_{1.5}MnNi_{0.5}$ alloy (figure 11) does not indicate a significant variation, with a slightly higher value for the high rate solidification process.

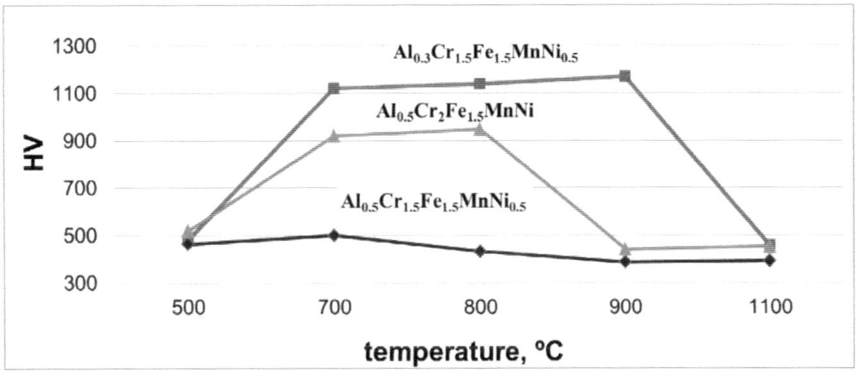

Figure 10. Microhardness variation with annealing temperature

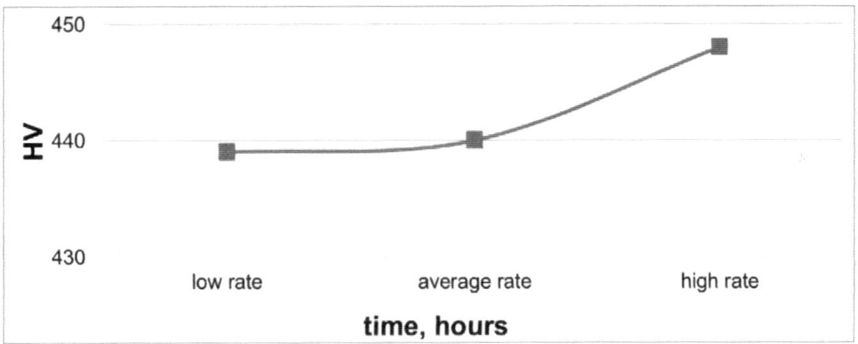

Figure 11. Microardness function of solidification rate for $Al_{0.3}Cr_{1.5}Fe_{1.5}MnNi_{0.5}$ alloy

3.3. Modelling results

Phase evolution and solidification of $Al_{0.5}Cr_{1.5}Fe_{1.5}MnNi_{0.5}$, $Al_{0.5}Cr_2Fe_{1.5}MnNi$ and $Al_{0.3}Cr_{1.5}Fe_{1.5}MnNi_{0.5}$ alloys were modelled with MatCalc software. MatCalc (The Materials Calculator) is a software toolbox for computational materials engineering that uses CALPHAD approach and covers the areas of multi-component phase equilibrium and thermodynamics as well as multi-phase precipitation kinetics and microstructure evolution in solid-state systems. Ni superalloy database was used for the present modelling goal. Nevertheless, the related diffusion database is not yet conclusive enough to be used in the precipitation and heat treatment calculations for high entropy alloys. Trials in this direction did not offer trusted results that could be used in our investigation.

The thermodynamic evaluation (figures 12-14) showed that all the alloys have a similar phase constitution based on sigma phase, BCC-B2 (NiAl), BCC-A2 and FCC-A1 types. The $Al_{0.5}Cr_{1.5}Fe_{1.5}MnNi_{0.5}$ has a smaller melting temperature than the other alloys, even if Al is in higher proportion. $Al_{0.5}Cr_{1.5}Fe_{1.5}MnNi_{0.5}$ alloy has a smaller FCC-A1 phase fraction than the other alloys, and in the $Al_{0.5}Cr_2Fe_{1.5}MnNi$ alloy remains stable until 1000ºC. The sigma phase is found to be dominant in the $Al_{0.3}Cr_{1.5}Fe_{1.5}MnNi_{0.5}$ alloy. The NiAl-B2 phase is found in similar proportion in all the alloys. The solidification behaviour, determined by Scheil-Gulliver method, is also similar for all the alloys.

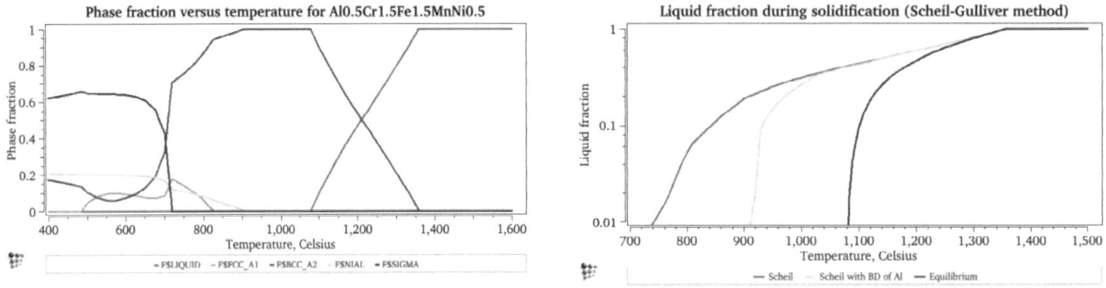

Figure 12. Phase equilibrium and solidification of $Al_{0.5}Cr_{1.5}Fe_{1.5}MnNi_{0.5}$ alloy

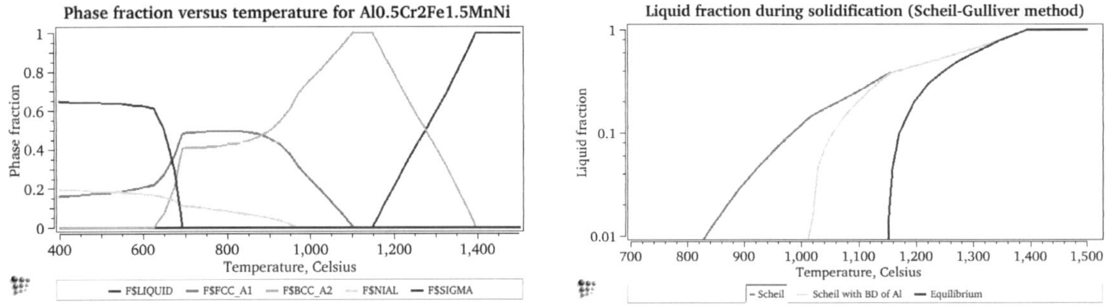

Figure 13. Phase equilibrium and solidification of $Al_{0.5}Cr_2Fe_{1.5}MnNi$ alloy

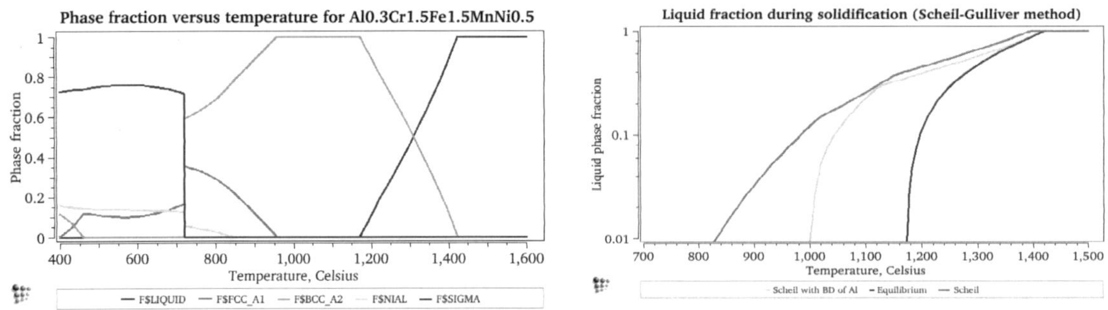

Figure 14. Phase equilibrium and solidification of $Al_{0.3}Cr_{1.5}Fe_{1.5}MnNi_{0.5}$ alloy.

Table 3. Phases found through the experimental and modelling approaches.

Alloy	as-cast	heat treated 700ºC	heat treated 800ºC	heat treated 900ºC	modelling
$Al_{0.5}Cr_{1.5}Fe_{1.5}MnNi_{0.5}$	BCC-A2 BCC-B2	σ phase BCC-B2	BCC-B2 σ phase	BCC-A2 BCC-B2 σ phase	σ phase BCC-B2(NiAl) BCC-A2
$Al_{0.5}Cr_2Fe_{1.5}MnNi$	BCC-A2 BCC-B2 FCC-A1	σ phase BCC-B2	σ phase FCC-A1 BCC-B2	BCC-A2 FCC-A1	σ phase BCC-B2(NiAl) BCC-A2
$Al_{0.3}Cr_{1.5}Fe_{1.5}MnNi_{0.5}$	BCC-A2 σ phase	σ phase BCC-A2 FCC-A1	σ phase FCC-A1 BCC-A2	σ phase FCC-A1 BCC-A2	σ phase BCC-B2(NiAl) BCC-A2

Calculations for phase stability at room temperature revealed structures formed of major sigma phase with minor B2-NiAl and BCC-A2 phases, for all the alloys. The sigma phase was determined to have molar fractions of: 0.59 for $Al_{0.5}Cr_{1.5}Fe_{1.5}MnNi_{0.5}$ alloy, 0.66 for $Al_{0.5}Cr_2Fe_{1.5}MnNi$ alloy and 0.7 for $Al_{0.3}Cr_{1.5}Fe_{1.5}MnNi_{0.5}$ alloy.

Table 3 contains the phases found in the modelling results and experimentally in all the alloys at as-cast and annealed states, in proportional order. The correlation of empirical and modelling results showed significant differences. In the as cast state only the $Al_{0.3}Cr_{1.5}Fe_{1.5}MnNi_{0.5}$ presented sigma phase and in low proportion. The $Al_{0.5}Cr_2Fe_{1.5}MnNi$ alloy contains a low proportion of FCC-A1 phase that was not predicted by the model. The annealing results enhanced sigma phase stabilisation in all the alloys. Still FCC-A1 phase is present in the $Al_{0.3}Cr_{1.5}Fe_{1.5}MnNi_{0.5}$ similar to the findings of other authors [17].

One has to take into account that in conventional alloys the annealed state at slow cooling rates shows similar phases in the as-cast state, which is not the case in present HEA. Longer annealing holding times (longer than 50 hours) or much slower solidification rates obtained by controlled casting techniques could provide more accurate results.

The criteria for the prediction of solid solution formation in high entropy alloys were partially validated by the experimental trials and proved conclusive in relation to the modelling results.

4. Conclusions

Selected Co-less high entropy alloy compositions from the Al-Cr-Fe-Mn-Ni system were studied to determine structural behaviour in as-cast and annealed state.

Criteria calculation for several alloy compositions predicted formation of mixed solid solution phases (BCC and FCC) and intermetallic structures.

$Al_{0.5}Cr_{1.5}Fe_{1.5}MnNi_{0.5}$, $Al_{0.5}Cr_2Fe_{1.5}MnNi$ and $Al_{0.3}Cr_{1.5}Fe_{1.5}MnNi_{0.5}$ alloys were further prepared by induction melting and annealed at various temperatures.

Optical microscopy revealed similar dendritic structures with an additional phase distinguished in the $Al_{0.3}Cr_{1.5}Fe_{1.5}MnNi_{0.5}$ alloy. Higher magnifications provided by SEM analyses revealed a uniform distribution of small precipitated phases. XRD determinations showed significant presence of BCC-A2 phase in all the alloys.

Lower solidification rates of the $Al_{0.3}Cr_{1.5}Fe_{1.5}MnNi_{0.5}$ alloy, analysed at lower magnifications, determined a significant increase in the interdendritic phase and a decrease in the hard phase presence.

SEM-EDS analysis of the heat treated $Al_{0.3}Cr_{1.5}Fe_{1.5}MnNi_{0.5}$ alloy showed similar uniform distribution of the precipitated phase but with larger precipitate sizes found in the samples heat treated at higher temperatures. Aluminum and nickel were found highly concentrated in the precipitated region.

Phase composition determined by x-ray analysis showed major sigma phase and BCC-B2 presence in all the alloys.

The microhardness values were strongly related to the proportion of sigma phase in the alloys, besides the $Al_{0.5}Cr_{1.5}Fe_{1.5}MnNi_{0.5}$ sample heat treated at 700ºC that presented unexpectedly low value.

Thermodynamic modelling of the selected high entropy alloys performed by the MatCalc software revealed suggestive results by predicting phase evolution for the specific compositions on a large temperature range, from 400ºC to 1500ºC. The study of alloys solidification by Scheil-Gulliver method was also provided. Precipitation and heat treatment results are not yet available for the selected high entropy alloys due to the lack in diffusion database research.

The correlation of experimental and modelling results revealed significant differences. Sigma phase was predicted by the modelling software as the main phase to form in all the alloys at room temperature, but was less found in the as-cast structures and not as predominant in the heat treated structures.

The initial theoretical criteria for phase prediction was entirely validated by the modelling method and partially validated by the experimental findings

Acknowledgements
The research work was made possible by the financial support of COST action CA15102 project, M-era.Net project no. 75/2017 and National project PNIII-134PED/2017.

References

[1] Yeh J W 2006 Annales de Chimie - Science des Matériaux volume 31 no 6 pp 633-648
[2] Guo S 2015 Mater. Sci. Technol. volume 31 no 10 pp. 1223–1230
[3] Troparevsky M C Morris J R Kent P R C Lupini A R Stocks G M 2015 Phys. Rev. X volume 5 issue 1 pp 1-6
[4] Senkov O N Miracle D B 2016 J. Alloys Compd. vol 658 pp 603–607
[5] Zhang C Zhang F Chen S Cao W 2012 JOM volume 64 pp 839.
[6] Zhang F Zhang C Chen S L Zhu J Cao W S Kattner U R 2014 Calphad: Computer Coupling of Phase Diagrams and Thermochemistry volume 45 pp 1–10.
[7] Widom M Huhn W P Maiti S and Steurer W 2014 Metall. Mater. Trans. A volume 45 pp 196.
[8] Middleburgh S C King D M Lumpkin G R Cortie M and Edwards L 2014 J. Alloys Compd. volume 599 pp 179–182.
[9] Ma D Grabowski B Körmann F Neugebauer J and Raabe D 2015 Acta Mater. volume 100 pp 90–97.
[10] Troparevsky M C Morris J R Daene M Wang Y Lupini A R and Stocks G M 2015 JOM volume 67 pp 2350–2363.
[11] Zhang T Zuo T Tang Z Gao M C Dahmen K A Liaw P K and Lu Z P 2013 Prog. Mater. Sci. volume 61 pp 1-93.
[12] Miracle D B Miller J D Senkov O N Woodward C Uchic M D and Tiley J 2014 Entropy volume 16 no 1 pp 494–525
[13] Miracle D B and Senkov O N 2017 Acta Mater. volume 122 pp 448–511
[14] Tung C C Yeh J W Shun T Chen S K Huang Y S and Chen H C 2007 Mater. Lett. volume 61 No. 1 pp 1-5
[15] Wang Y P Li B S Ren M X Yang C and Fu H Z 2008 Mater. Sci. Eng. A volume 491 No 1-2 pp 154-158
[16] Wen L H Kou H C Li J S Chang H Xue X Y and Zhou L 2009 Intermetallics volume 17 No 4 pp 266-269.
[17] Chen S T Tang W Y Kuo Y F Chen S Y Tsau C H Shun T T Yeh J W 2010 Mater. Sci. Eng. A volume 527 No 21-22 pp 5818-5825
[18] Guo S Hu Q Ng C and Liu C T 2013 Intermetallics volume 41 pp 96–103
[19] Wang Z Guo S and Liu C T 2014 JOM volume 66 issue 10 pp 1966-1972
[20] Zhang Y Zhou Y J Lin J P Chen G L and Liaw P K 2008 Adv. Eng. Mater. volume 10 pp 534–538
[21] Guo S and Liu C T 2011 Prog. Nat. Sci. Mater. Int. volume 21 pp 433–446.
[22] Yang X and Zhang Y 2012 Mater. Chem. Phys. volume 132 No 2-3 pp 233–238.
[23] Poletti M G and Battezzati L 2014 Acta Mater. volume 2014 75 pp 297–306.
[24] Guo S Ng C Lu J and Liu C T 2011 J. Appl. Phys. volume 109 pp 1–5
[25] Tsai M H Chang K C Li J H Tsai R C and Cheng A H 2016 Mater. Res. Lett. volume 4 no 2 pp 1–6
[26] Tsai M H Li J H Fan A C and Tsai P H 2017 Scr. Mater. volume 127 pp 6–9

Optimization of the A-TIG welding for stainless steels

M Jurica, Z Kožuh, I Garašić and M Bušić

University of Zagreb, Faculty of Mechanical Engineering and Naval Architecture, Ivana Lučića 5, 10002 Zagreb, Republic of Croatia

E-mail: matija.busic@fsb.hr

Abstract. The paper presents the influence of the activation flux and shielding gas on tungsten inert gas (A-TIG) welding of the stainless steel. In introduction part, duplex stainless steel was analysed. The A-TIG process was explained and the possibility of welding stainless steels using the A-TIG process to maximize productivity and the cost-effectiveness of welded structures was presented. In the experimental part duplex, 7 mm thick stainless steel has been welded in butt joint. The influence of activation flux chemical composition upon the weld penetration has been investigated prior the welding. The welding process was performed by a robot with TIG equipment. With selected A-TIG welding technology preparation of plates and consumption of filler material (containing Cr, Ni and Mn) have been avoided. Specimens sectioned from the produced welds have been subjected to tensile strength test, macrostructure analysis and corrosion resistance analysis. The results have confirmed that this type of stainless steel can be welded without edge preparation and addition of filler material containing critical raw materials as Cr, Ni and Mn when the following welding parameters are set: current 200 A, welding speed 9,1 cm/min, heat input 1,2 kJ/mm and specific activation flux is used.

1. Introduction

Stainless steel is the name given to a group of corrosion-resistant steels. Stainless steels have a chromium content of at least 10,5 %. The high degree of chromium activity is actually the principal basis for utilizing it as an alloying element in corrosion resisting alloys. As a result of reactions of chromium with the oxygen from the air, a protective oxide film forms on this alloy and prevents further rapid oxidation. Varying additions of Ni, Mo, N, Cu, Mn, W, Ti, Ni and other elements may also be present. [1-6]. Stainless steels are classified on the basis of their matrices as ferritic, martensitic and austenitic stainless steels. Even in the fully soft condition stainless steels have tensile strengths greater than that of mild steel. The soft austenitic steels have exceptional ductility, elongation exceeding 50 %. The austenitic stainless steels can therefore be cold worked to give an extensive range of semi-fabricated and finished products. The cold working may be adjusted so that the final condition attained has the most advantageous combination of strength and toughness. The ferritic stainless steels offer good strength and ductility, but without the outstanding formability of the austenitic varieties.

Duplex stainless steels (DSSs) with austenitic and ferritic grains possess beneficial combinations of these two phases. The ferrite/austenite ratio in DSS must be close to 50:50 by adding appropriate alloying elements for achieving desired microstructures and mechanical properties. DSS exhibit greater toughness and better weldability than ferritic stainless steel [5]. Compared with austenitic grades, DSS have higher resistance to pitting and stress corrosion cracking. Accordingly, they are widely used in various chemical, petrochemical, food, power, transportation, paper and oil industries. Most of these applications require welding as a joining method for parts made from DSS. DSSs undergo microstructural changes during heat treatment, welding process and prolonged engineering service at high temperatures [5]. In the welding of DSS, it is essential to maintain a ferrite–austenite ratio close to 50:50. This phase balance, may however, be upset due to rapid cooling involved in most weld thermal cycles resulting in weld metal ferrite contents in excess of 50 %. The resultant phase ratio is dependent on the energy input during welding, as this determines the cooling rate and the extent of the phase transformation which is diffusion based. If high heat inputs are used, coarse grains are produced in the weld region, wide heat-affected zones and possibly, precipitation of brittle intermetallic phases may develop [6].

Tungsten inert gas (TIG) welding is a widely used technology for joining stainless steels. When conventional TIG welding process is applied for thick sections it has some disadvantages as low penetration and small deposit, and consequently reduced productivity compared to the other welding processes. Material thickness greater than 3 mm is necessary to weld in multiple passes and this greatly reduces productivity of the process.

One variety of TIG is welding with activation flux coating (A-TIG) on the upper surface of the plates that should be joined. Activating flux is a mixture of inorganic material suspended in an evaporative medium. A thin layer of flux is applied on the surface of the joint before welding [7]. One interpretation is that activating flux influences on the penetration with changing the surface tension in the weld pool so that the fluid flow is opposite of conventional TIG, and tends to increase the penetration [7].

International standard ISO 9692-1:2013 “Welding and allied processes - Types of joint preparation - Part 1: Manual metal-arc welding, gas-shielded metal-arc welding, gas welding, TIG welding and beam welding of steels” gives a recommendation for edge preparation according to the thicknesses of parts to be welded (figure 1). According to this standard for TIG welding of steels in thicknesses $3 < t \leq 8$ mm edge preparation should be square groove with gap between plates $b \approx t$, where t is the thickness of the plate (figure 1.a). Edge preparation for thicknesses $3 < t \leq 10$ should be single V groove with root gap $b \leq 4$ and a V groove angle $40° \leq \alpha \leq 60°$ (figure 1.b). It is notable that edge preparation requires material loss prior the welding and additional filler material during the welding. A-TIG technology makes it possible to intensify the conventional TIG practices for joining the thickness of 8 - 10 mm by single pass full penetration welds, with no edge preparation [7]. In A-TIG welding penetration is increased with the application of a thin coating of activating flux material on the work piece surface prior to welding. The effect of flux is to constrict the arc which increases the current density at the anode root and the arc force on the weld pool. With A-TIG technology the edge preparation is simplified, loss of the material and additional filler material are avoided.

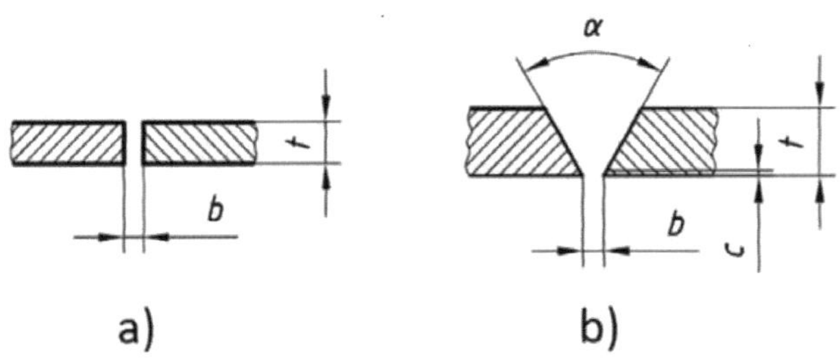

Figure 1. Joint preparation for TIG welding of steels according to the ISO 9692-1:2013

General objective of this research was to clarify the improvements in productivity when TIG welding is applied with activation flux on austenitic-ferrite (duplex) 7 mm thick stainless steel. In the first part specific type of the activation flux has been prepared and the influence on the penetration in bead on plate welding has been examined. According to the achieved penetration parameters for welding two butt joints have been produced with variable shielding gases. Welded joints have been subjected to visual control and destructive tests to verify the validity of the welding parameters.

2. Experimental

2.1 Material

Standard duplex stainless steel with designation X2CrNiMoN22-5-3 has been used in this experiment. Mechanical properties of this material according to the standard HRN EN 10088-2:2015 [8] are presented in table 1. Chemical composition of the same material has been analysed with Energy-dispersive X-ray spectroscopy (EDS) on scanning electron microscope Tescan Vega TS5136 [9]. The results are presented in table 2. The X2CrNiMoN22-5-3 is austenitic-ferritic stainless steel with:

a) high resistance to stress corrosion cracking (SCC) in chloride-bearing environments and environments containing hydrogen sulphide;
b) high resistance to general corrosion, pitting and crevice corrosion;
c) high resistance to erosion corrosion and corrosion fatigue;
d) high mechanical strength - roughly twice the proof strength of austenitic stainless steels;
e) physical properties that offer design advantages;
f) good weldability.

Activation flux has been prepared from powders SiO_2 and Cr_2O_3 in proportion 1:1. In order to prepare highly concentrated powder suspensions 96 % ethyl alcohol has been added in proportion 1:2 to this mixture.

Table 1. Mechanical properties of duplex stainless steel X2CrNiMoN22-5-3 [8]

Thickness [mm]	Yield stress min. $R_{p0,2}$ [MPa]	Tensile strength R_m [MPa]	Elongation min. A [%]
7	460	640 to 840	25

Table 2. Chemical composition of duplex stainless steel X2CrNiMoN22-5-3 [9]

C	Mn	P	S	Si	Ni	Cr	Mo	N
0,014	1,903	0,026	0,002	0,278	5,005	22,421	2,559	0,1690

2.2 Penetration of the electric arc

Penetration of the electric arc has been tested on the test coupons in dimensions 200 × 200 × 7 mm. Plates were cleaned with acetone prior deposition of the activation flux. One thin layer of activation flux was laid on the surface of the work-piece using a paintbrush prior each A-TIG bead on plate welding. Welding was performed with FRONIUS MagicWave 2200 Job G/F TIG welding machine. Thoriated Tungsten electrode with diameter 2,4 mm and designation WT20 has been used for welding. In order to have a linear motion of the electric arc with accurate speed of moving a TIG welding torch has been installed on the „BUG-O Systems" Modular Drive Systems automatic machine used for guidance in welding. Two types of shielding gas have been used with the same flow of 9 l/min. First of them was 100 % Argon 4.8 (with a purity of 99,998 %) and the other one was a mixture of 97,5 % Argon with 2,5% Nitrogen. Nitrogen has been added to stimulate occurrence of austenitic phase and prevent ferritization. The electrode tip workpiece distance was constantly at 3 mm. Table 3 presents welding parameters for all weld runs in the first part of the experiment. Heat input is defined with the equation:

$$Q = \frac{k \times U \times I \times 60}{v \times 10000} \ \text{[kJ/mm]}$$

where U is voltage, I is current, v is welding speed and k is the coefficient of thermal efficiency of the specific welding process (for TIG it is 0,6).

IOP Conf. Series: Materials Science and Engineering **329** (2018) 012012 doi:10.1088/1757-899X/329/1/012012

Table 3. Welding parameters in bead on plate welding [9]

Sample	Activation flux	Shielding gas	Welding speed [cm/min]	Current [A]	Voltage [V]	Heat input [kJ/mm]
1A	NO	100 % Ar	10,2	160	11,8	0,666
2A	NO	100 % Ar	7,8	200	13,1	1,209
3A	YES	100 % Ar	10,2	160	13,5	0,762
4A	YES	100 % Ar	7,8	200	14,2	1,310
1B	NO	Ar + 2,5 % N_2	10,2	160	13,6	0,768
2B	NO	Ar + 2,5 % N_2	7,8	200	14,3	1,320
3B	YES	Ar + 2,5 % N_2	10,2	160	13,7	0,774
4B	YES	Ar + 2,5 % N_2	7,8	200	14,8	1,366

2.3 A-TIG welding parameters

After welding and examination of bead on plate welds, parameters for welded joint have been determined. The constant parameters were:

- welding speed 9,1 cm/min;
- shielding gas flow 9 l/min;
- electrode tip workpiece distance 3 mm;
- tungsten electrode diameter 2,4 mm;
- welding torch inclination 90°.

Two welded joints have been produced without any filler material, first one with 100 % Ar shielding gas (marked as Z1) and the other one with Ar + 2,5 % N_2 shielding gas (marked as Z2). Plates in dimensions 200 x 100 x 7 mm were placed on a 10 mm thick cooper backing plate. Plates had single square edge preparation without any spacing between. Samples have been welded with application of the same activation flux in width of 20 mm on the upper surface. Table 4. Presents welding parameters used for producing welded joints.

Table 4. Welding parameters used for producing welded joints [9]

Sample	Activation flux	Shielding gas	Welding speed [cm/min]	Current [A]	Voltage [V]	Heat input [kJ/mm]
Z1	YES	100 % Ar	9,1	200	14,7	1,163
Z2	YES	Ar + 2,5 % N_2	9,1	200	15,4	1,218

2.4 Preparation of samples for analysis

From the produced welds specimens for tensile strength testing, and macrostructure analysis have been sectioned perpendicular to the welding direction using a circular band saw. Specimens for tensile strength test have been machined to required dimensions using a universal milling machine according to the requirements of the standard HRN EN ISO 4136:2013.

Specimens for analysis of macrostructure have been grinded and polished using sandpapers: 1) P320, 2) P500, 3) P1000, 4) P2000 and 5) P4000. After polishing the specimens were etched for 40 s in reagent composed of NaOH and H_2O in proportion 2:3 and using DC current with voltage 2 V.

Specimens for corrosion resistance analysis according to standard ASTM G48-03 (method – A) have been plasma cut in dimensions 100 × 45 mm and washed with hot water to eliminate grease and impurities from the surface.

3 Results

3.1 Visual control

During welding with active flux, the weld pool was calm as it is characteristic for the TIG welding process. There were no spatters on the surface of the material around the produced welds. Sample 1A has a wider bead on the plate weld in comparison with sample 3A. Heat affected zone and the root of the weld are wider on the sample 3A in comparison with sample 1A. Welds produced with higher current (samples 2A and 4A) have the same influence of the A-TIG welding on the width of the bead on the plate weld. Sample 4A has sufficient penetration on the root side and narrower bead on plate width in comparison with sample 2A that is without root penetration. The difference between samples made with the second shielding gas (Ar + 2,5 % N_2) is the same as with 100 % Ar.

Figure 2 presents weld face and weld root on Z1 specimen. Figure 3 presents weld face and weld root on Z2 specimen. Appearance of the weld on the face side on both samples is very similar and depends only on the guidance of the electric arc. At the root side of the specimen Z1 lack of fusion can be observed after a few centimetres of a correct joint. The reason can be in the incorrect guidance of the electric arc after half of the path. At the root side of the specimen Z2 full penetration can be observed.

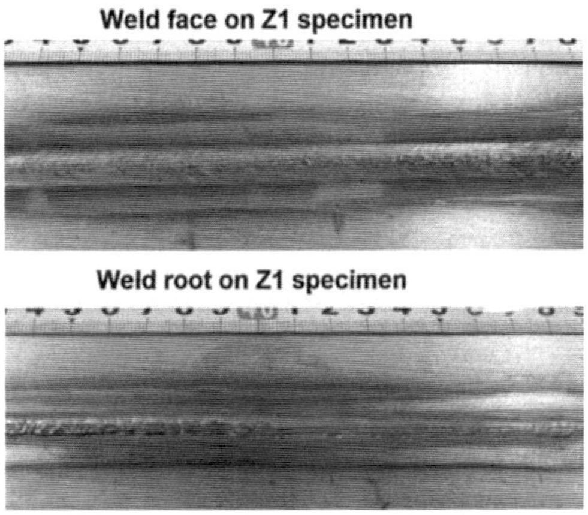

Figure 2. Weld face and weld root on Z1 specimen [9]

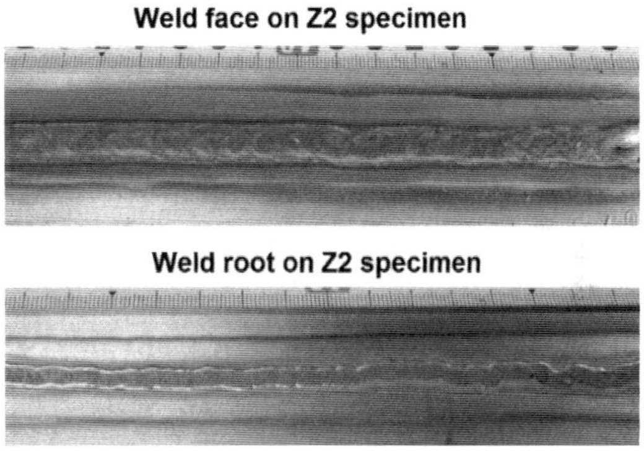

Figure 3. Weld face and weld root on Z2 specimen [9]

3.2 Macrostructure analysis

Macrostructure has been examined on a LEICA MZ6 stereomicroscope, using Leica Application suite version 3.2.0 software. Figure 4 presents a comparison of geometrical features in the produced bead on plate welds in the first part of the experiment. Samples made with 100 % Ar shielding gas are located on the left and samples made with Ar + 2,5 % N_2 are located on the right side of the figure 4. The difference of the geometrical features in dependence on the shielding gas type is clearly seen. When Ar + 2,5 % N_2 has been used, the produced bead on plate welds have wider and shallower penetration in comparison with 100 % Ar shielding gas. Sufficient penetration through the whole thickness has only been achieved when welding current was 200 A and welding speed was 7,8 cm/min.

Figure 5 presents geometrical features of the produced welds Z1 and Z2. Both welds have a minimal face and root reinforcement but have proper shape. Specimen Z1 has a face width 8,1 mm and root reinforcement 0,3 mm. Specimen Z2 made with Ar + 2,5 % N_2 has a smaller face width, 7,6 mm and the root reinforcement 0,4 mm. Transition from an elongated base metal microstructure (obtained with forming of the sheets) to the weld metal microstructure is clearly seen on both welds. It is also possible to see molten metal flow contours from the outside to the inside. This feature is a clear footprint of the Marangoni's effect.

IOP Conf. Series: Materials Science and Engineering **329** (2018) 012012 doi:10.1088/1757-899X/329/1/012012

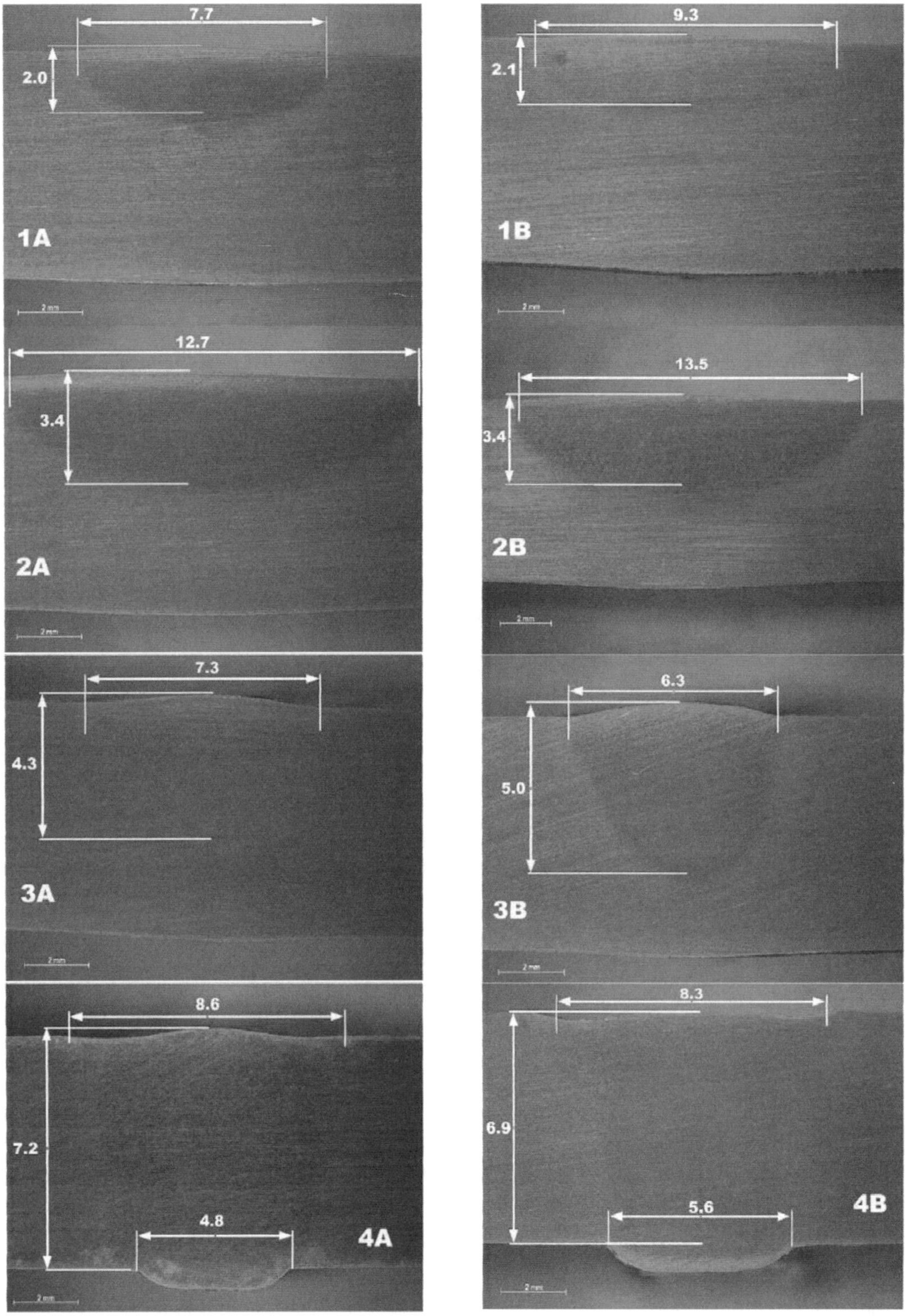

Figure 4. Comparison of geometrical features in the produced bead on plate welds

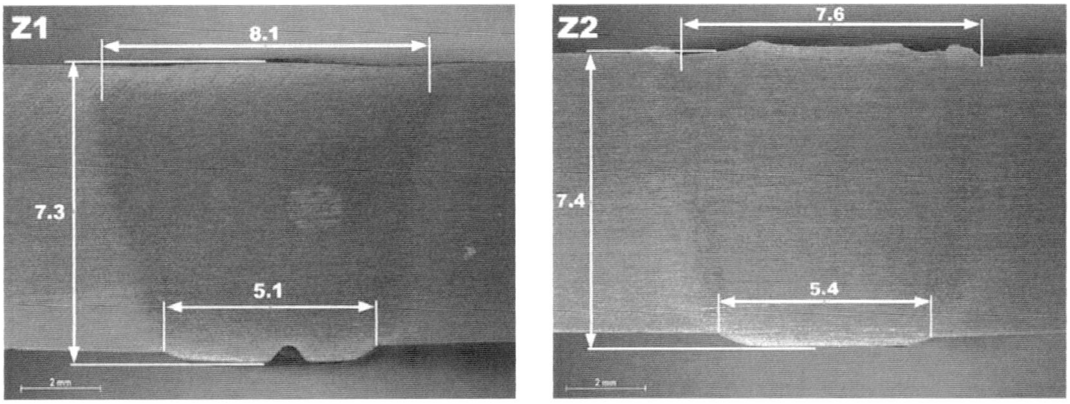

Figure 5. Geometrical features of the produced welds Z1 and Z2

3.3 Tensile strength testing

The uniaxial tensile testing of the specimens has been conducted according to HRN EN ISO 4136:2013 standard [10]. Figure 6 presents samples after the tensile strength testing. Sample Z1 welded with 100 % Ar shielding gas has a fracture in the heat affected zone. Sample Z2 welded with mixture shielding gas 2,5 % N_2 and 97,5 % Ar has a fracture in the base material approx. 40 mm far from the weld centre. Table 5 presents results of the transversal tensile strength testing of the produced welds. From the measured results (table 5) it can be clear that both welds have similar mechanical properties. Measured mechanical properties of the produced welds also fulfil requirements of the HRN EN 10088-2:2015 [8] standard for the base material.

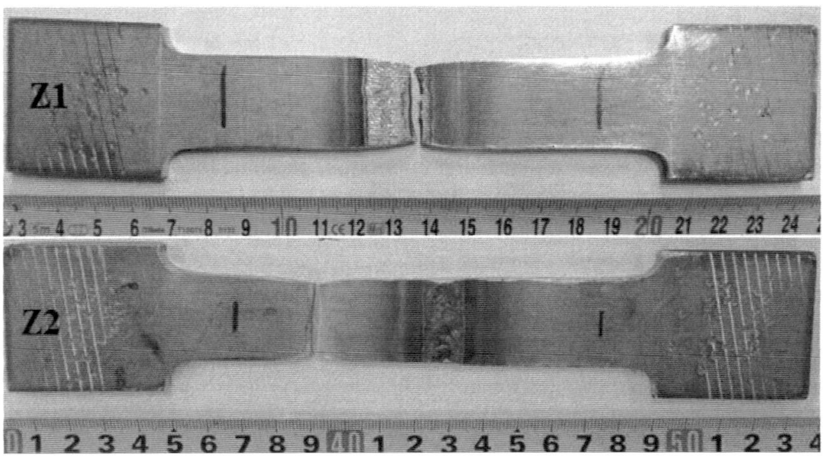

Figure 6. Testing samples after transversal tensile testing [9]

Table 5. Results of the transversal tensile strength testing of the produced welds [9]

Sample	R_m [N/mm^2]	F_m [N/mm^2]	$R_{p0,2}$[N/mm^2]	F_e [kN]	S_0[mm^2]	F_b [kN]
Z1	773,23	136,63	679,12	120	176,70	70,27
Z2	776,45	138,65	683,21	122	178,57	85,29

3.4 Corrosion resistance analysis

Corrosion resistance analysis of the produced welds has been conducted according to the standard ASTM G48 - 03 "Standard test methods for pitting and crevice corrosion resistance of stainless steels and related alloys by the use of ferric chloride solution "(method A - Ferric chloride pitting test) [11]. Samples were immersed in 10% solution of $FeCl_3$ in H_2O with a temperature of 50°C for a period of 48 h. Figure 7 presents samples after corrosion analysis. It can be observed that most of the pits have been developed in the weld metal on samples 1A, 3A and 4A. Weld metal on sample 2A is free of any pits. Samples welded with mixture shielding gas 2,5% N_2 and 97,5% Ar (1B, 2B, 3B and 4B) are free of pits in the weld metal and the heat affected zone. However, some pits occurred in the base material a few millimeters away from the heat affected zone. A-TIG welding resulted with less pits (3B and 4B) in comparison with conventional TIG welding (1B and 2B).

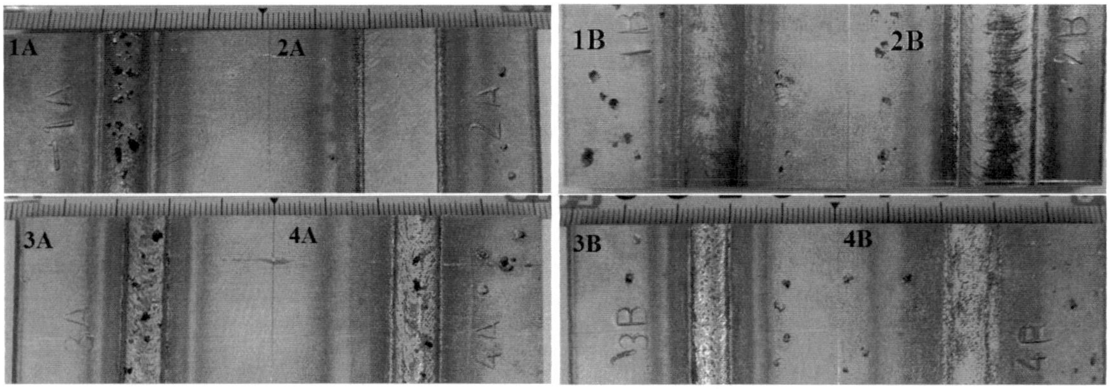

Figure 7. The appearance of bead on plate welds after 48 hours exposure in aggressive solution [9]

4 Conclusions

From the obtained results in this investigation the following conclusions can be summarized:

a) When comparing bead on plate width A-TIG welding produces narrower welds in comparison with conventional TIG. The same influence is observed when welding with current 160 A and with 200 A.
b) Appearance of the weld on the face side of samples welded with A-TIG and conventional TIG welding is almost the same and depends only on the guidance of the electric arc.
c) Macrostructure analysis of geometrical features of the produced bead on plate welds has confirmed that A-TIG welding intensifies the penetration and produces narrower and deeper welds. Shielding gasses used in this experiment have small influence on the geometrical features of the produced welded joints. Full penetration has been achieved only in A-TIG welding with 200 A and with welding speed 7,8 cm/min.
d) Transversal tensile testing of the produced welds has confirmed that both shielding gases influence similar on the tensile strength. Measured mechanical properties of the produced welds fulfill requirements of the standard EN 10088-2:2015 for the base material.
e) Corrosion resistance analysis of the produced welds has confirmed a positive effect of mixture shielding gas 2,5 % N_2 and 97,5 % Ar (1B, 2B, 3B and 4B) on increasing resistance to pitting and crevice corrosion. Bead on plate welds made with mixture shielding gas were free of any pits in weld metal and heat affected zone. A-TIG bead on plate welding resulted with less pits (samples 3B and 4B) in comparison with conventional TIG welding (samples 1B and 2B).
f) A-TIG welding is fully applicable technology for welding butt joint of duplex stainless steel in thickness 7 mm, with mixture shielding gas 2,5 % N_2 and 97,5 % Ar, welding speed and current and other constant parameters according to this experiment.

References

[1] Umoru L E, Afonja A A and Ademodi B Corrosion study of AISI 304, AISI 321 and AISI 430 stainless steels in a tar sand digester, *Journal of Minerals & Materials Characterization &*

Engineering, vol. 7, no.4, pp 291-299, 2008.
[2] Otokumpu Stainless AB *Handbook of Stainless Steel*, Avesta Resarch Centre, Avesta, Sweden, 2013.
[3] Davis J R *Stainless steels*; ASM International, USA, 1994.
[4] Gun R N *Duplex stainless steels: microstructure, properties and applications*; Woodhead Publishing: Cambridge, UK, 1997.
[5] Chan K W and Tjong S C Effect of secondary phase precipitation on the corrosion behavior of duplex stainless steels, *Materials*, no. 7, pp. 5268-5304, 2014.
[6] Gideon B, Ward L and Biddle G Duplex stainless steel welds and their susceptibility to intergranular corrosion, *Journal of Minerals & Materials Characterization & Engineering*, vol. 7, no.3, pp 247-263, 2008.
[7] Shyu S W, Huang H Y, Tseng K H and Chou C P Study of the performance of stainless steel A-TIG welds, *Journal of Materials Engineering and Performance*, vol. 17, no. 2, pp 193-201, 2008.
[8] HRN EN 10088-2:2015, Stainless steels -- Part 2: Technical delivery conditions for sheet/plate and strip of corrosion resisting steels for general purposes
[9] Babić N *A-TIG welding of duplex steels*, Diploma thesis, University of Zagreb, Faculty of Mechanical Engineering and Naval Architecture, 2016.
[10] HRN EN ISO 4136:2013, Destructive tests on welds in metallic materials – Transverse tensile test
[11] ASTM G48 – 03 - Standard test methods for pitting and crevice corrosion resistance of stainless steels and related alloys by use of ferric chloride solution

Intermetallics as innovative CRM-free materials

Pavel Novák[1], Lucyna Jaworska[2], Marcello Cabibbo[3]

[1] Department of Metals and Corrosion Engineering, University of Chemistry and Technology Prague, Technická 5, 16628 Prague. Czech Republic
[2] The Institute of Advanced Manufacturing Technology, Wroclawska 37A, 30-011 Krakow, Poland
[3] DIISM/Università Politecnica delle Marche, Via Brecce Bianche, 60131-Ancona, Italy
panovak@vscht.cz

Abstract. Many of currently used technical materials cannot be imagined without the use of critical raw materials. They require chromium (e.g. in stainless and tool steels), tungsten and cobalt (tool materials, heat resistant alloys), niobium (steels and modern biomaterials). Therefore there is a need to find substitutes to help the European economy. A promising solution can be the application of intermetallics. These materials offer wide variety of interesting properties, such as high hardness and wear resistance or high chemical resistance. In this paper, the overview of possible substitute materials among intermetallics is presented. Intermetallics based on aluminides and silicides are shown as corrosion resistant materials, composites composed of ceramics in intermetallic matrix as possible tool materials. The manufacturing processes are being developed to minimize the disadvantages of these materials, mainly the room-temperature brittleness.

1. Introduction

Metallic materials are used in structural applications for their typical properties, such as good strength and toughness at common temperatures, ductility, electric and thermal conductivity etc. In addition to this common application range, the use at the limits of these materials is also often required, e.g. at high temperatures, in strongly aggressive corrosion environment or under extreme abrasive or adhesive wear. High temperatures are very problematic for common metallic materials due to oxidation and gradual degradation by creep. For high-temperature application, iron-based alloys (i.e. heat resistant steels), nickel and cobalt alloys are commonly applied. However, these materials are characterized by relatively high density (7.8 - 9 g.cm-3), which is strongly disadvantageous e.g. for aerospace industry. Another important problem for EU economy is the fact that these materials contain elements (e.g Co, W), which are listed by European Commission as critical raw materials (CRM) [1]. Similar problem is in the case of tool materials, which contain Co and W as the main constituents (cemented carbides) and also elements with high economic risk just below the boundary defining CRMs (Cr, Mo). In addition, high resistance against mechanical wear is achieved by heat treatment in the case of tool steels and therefore the operating temperature is limited by the tempering temperature. Materials based on intermetallics could be promising substitutes of these materials.

Intermediate phases are formed in many systems of two or more metals or metals with non-metals when the solubility in metal-based solid solution is exceeded. Intermediate phases are chemical

compounds whose structure differs from all elements forming them. If they contain metals only, they are called intermetallic phases or intermetallic compounds, abbreviated as intermetallics. Compounds of metal with semimetal (silicides) and ordered solid solutions of aluminium in transition metals (aluminides, e.g. FeAl, Fe_3Al) are also usually denoted as intermetallics.
This paper aims to present the possible outlooks in application of intermetallic compounds as CRM substitutes.

2. Heat and corrosion resistant materials

The application as high temperature materials is probably dominant in the case of intermetallics. High melting points (e.g. 2130 °C for Ti_5Si_3 [2]), good creep properties and high-temperature oxidation resistance predestine them to this application range. Typical intermetallics for high temperatures are silicides (e.g. $MoSi_2$) for heating elements of electric resistance furnaces and nickel aluminides used mainly in the form of protective coatings for turbine blades for power plants. In addition to these materials, Ti-Al phases (mainly Ti_3Al a TiAl) and the ordered Fe_3Al a FeAl phases play an important role.
During the development of intermetallics based on Fe-Al and Ti-Al systems, the positive effect of Cr, W and Nb has been proved and therefore current commercial variants of these materials used e.g. in airplane jet engines are commercially produced [3]. However, these elements are listed as CRMs [1]. It has been also proved that the addition of silicon decreases the oxidation rate of Ti-Al alloys at the temperatures up to 1000 °C (Figure 1). Similarly, highly positive effect of silicon was found in the case of Fe-Al alloys. In both of these groups of materials, the silicon addition leads to the improvement of the protective properties of the oxide layer and also to the formation of silicides and other oxidation-resistant silicon-rich phases under the oxide layer (Figure 2) [4]. This sub-layer acts as a secondary barrier against oxygen penetration to the material, thus lowering the oxidation rate and protecting the core of the material. Due to these findings, further research was focused on Ti-Al-Si and Fe-Al-Si alloys. Silicon use can be managed even though it is also listed as CRM. As a source of this element, the secondary silicon from recycled electronics can be applied, because lower purity is required in these alloys. In addition, the silicon is listed as CRM probably mainly due to the lack of manufacturing capacities (silicon production plants) in Europe, even though there are relatively large reserves of the raw material (silica).

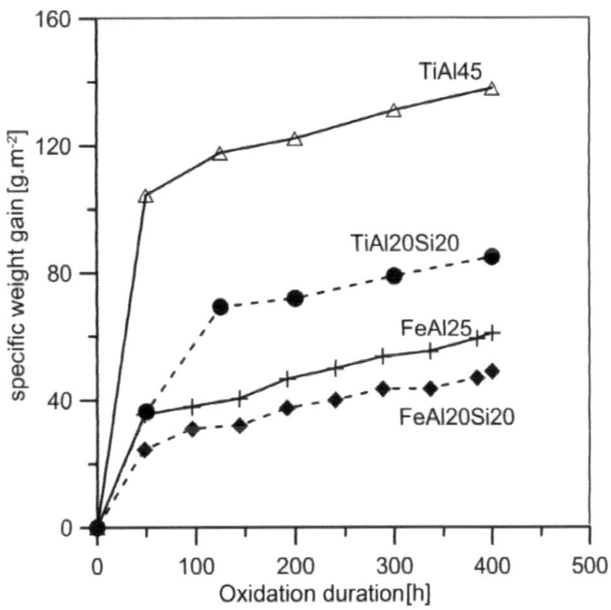

Figure 1: Oxidation resistance of Fe-Al-Si and Ti-Al-Si alloys at 800°C in air

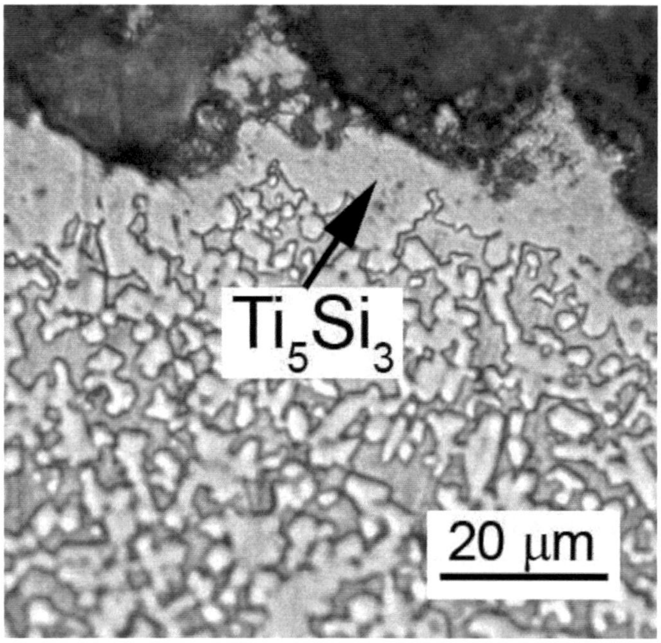

Figure 2: Microstructure of the near-surface are of the Ti-Al-Si alloy after oxidation [4]

Systematic testing of the influence of the other alloying elements on the properties of these materials led to the development of Fe-Ni-Al-Si alloy, which exhibits much lower density than common high-temperature materials (approx. 6 $g.cm^{-3}$) together with excellent oxidation resistance, thermal stability and relatively low price of the constituents [5].
It has been proved that the iron and titanium aluminides are also corrosion resistant in electrolytes [6]. Iron aluminide (FeAl) passivates by the Al_2O_3 layer in the solutions with pH above 3. Similar situation is for titanium aluminide, where this limit is at the pH of 2.5 - 3. When the material is alloyed by silicon, it is able to passivate by a layer of silicon oxide even below these values, but the passive layer is weaker.

3. Tool materials

The above mentioned aluminium and silicon alloyed materials exhibit also high wear resistance [7]. These properties predispose this alloy not only to the highly demanding high-temperature applications, but probably also for e.g.the high-speed machining tools, where high wear resistance and thermal stability are required. Other materials, which are the candidates for severe wear conditions as in tools are the composite materials with intermetallic matrix and ceramic reinforcement. One example of this type of materials is the composite with NiAl matrix reinforced by particles or short fibres of Al_2O_3 (Figure 3). This composite was successfully prepared by three powder metallurgy methods – reactive sintering of the mixture of nickel and titanium powders with alumina particles or short fibres [8], aluminothermic reaction between nickel oxide and aluminium [8], mechanical alloying of the mixture of nickel and aluminium with the addition of alumina particles and subsequent compaction by spark plasma sintering.

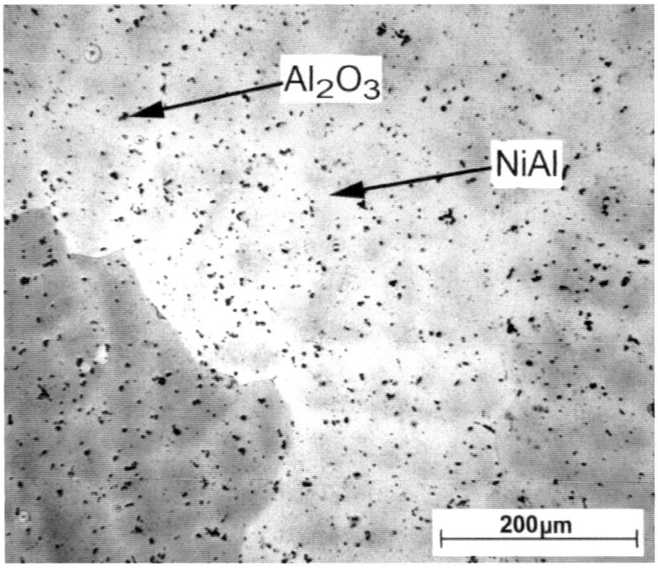

Figure 3: Microstructure of the NiAl-Al_2O_3 composite [8]

4. Processing of intermetallics

In processing of intermetallics, the casting technology is dominant, even more than in the case of conventional metallic materials. The reason is limited plasticity even at higher temperatures, which makes the forming processes feasible only in individual cases [9]. Ti-Al-Si and Fe-Al-Si alloys manufactured by melting and casting contain coarse, randomly oriented sharp-edged particles rich in silicon (Figure 4a). In the structure of Ti-Al-Si alloy this particle can be identified as Ti_5Si_3 silicide, while in Fe-Al-Si alloys a large series of various silicon-rich phases can be seen depending on the silicon content. There are FeSi and Fe_3Si silicides or Fe-Al-Si ternary phases (e.g. Al_2FeSi, $Al_2Fe_3Si_3$ or Al_2FeSi_3). These intermetallics are highly oxidation resistant, but also very brittle. This fact leads to undesirable mechanical properties.

Therefore, the processes for modification of the morphology of these phases were searched in order to eliminate this negative influence. The first approach was the experiment with the orientation of the Ti_5Si_3 particles, aiming to produce in-situ composite by elongated Ti_5Si_3 particles (fibres) in tougher matrix composed of Ti_3Al or TiAl phases. Directional solidification experiments were carried out in floating zone optical furnace in cooperation with the Institute of Physics AS CR. However, there were cracks in silicide fibres observed in the directionally solidified structure (Figure 4b). The reason for this phenomenon is different thermal expansion coefficient of Ti_5Si_3 silicide in dependence on crystal direction. Due to this fact, the stress is formed arising in the initiation of cracks. It strongly limits the applicability of Ti-Al-Si alloys produced by directional solidification [10].

Following research was focused on refinement of silicon-rich phases by powder metallurgy. Powder metallurgy processes used in production of intermetallics can be divided to two main groups. First one comprises the processes using the alloyed powders, i.e. already produced intermetallics. Alloyed powders can be prepared by milling of compact materials prepared by casting, melt atomization or by direct milling of elemental powders - by mechanical alloying [11]. The latter technology is able to prepare nano-grained powders (Figure 4c). However, this process is highly energy consuming. The next disadvantage of this technology is problematic densification of the powders of intermetallics. These powders are characterized by problematic compressibility and sinterability and therefore the consolidation requires application of special techniques, such as Hot Isostatic Pressing (HIP) or Spark Plasma Sintering (SPS). SPS technology is a modern process which uses compression and simultaneous passage of high electric current (e.g. 4000 V). It leads to rapid heating of the material and formation of discharge between the powder particles, resulting in fact in local welding of the powders to form compact material [12]. In order to lower the porosity of the product and to improve the mechanical properties correspondingly, the High Pressure Spark Plasma Sintering (HP SPS) has been developed and also already applied to intermetallics [13].

Second group of the powder metallurgy methods in production of intermetallics comprises the reaction synthesis processes using the chemical reactions between the powders of pure metals or alloys. In these processes called reaction sintering, the intermetallics are formed by thermally activated reactions between metallic powders. The initiation proceeds usually at the temperatures significantly lower than the melting point of prepared compound. Therefore the chemical reactions leading to the formation of intermetallics are usually strongly exothermic and therefore it is not needed to supply the heat after initiation and thus the reactions propagate by its own evolved energy. Due to this fact, this technology is often called SHS - Self-propagating High-temperature Synthesis [14]. Especially suitable seem to be the systems, where one reactant (e.g. aluminium) is molten before initiation. It fills the pores by capillary forces and decreases the porosity of the product. Nevertheless in the case of intermetallics in Fe-Al and Ti-Al systems reach the porosity above 25 vol. % [14,15]. The phenomenon is associated predominantly with significantly different diffusion rate of aluminium and iron or titanium, where the unidirectional diffusion is compensated by the diffusion of vacancies, and also with the temporary formation of phases with totally different structure (e.g. Fe_2Al_5). It leads to volume changes causing the porosity [16].

The research carried out in our department showed that the ternary alloys with silicon do not suffer from this problem. The reasons are in different mechanism of the reactions and also in high thermal effects in presence of silicon. It can cause local melting of the reaction mixture [17].

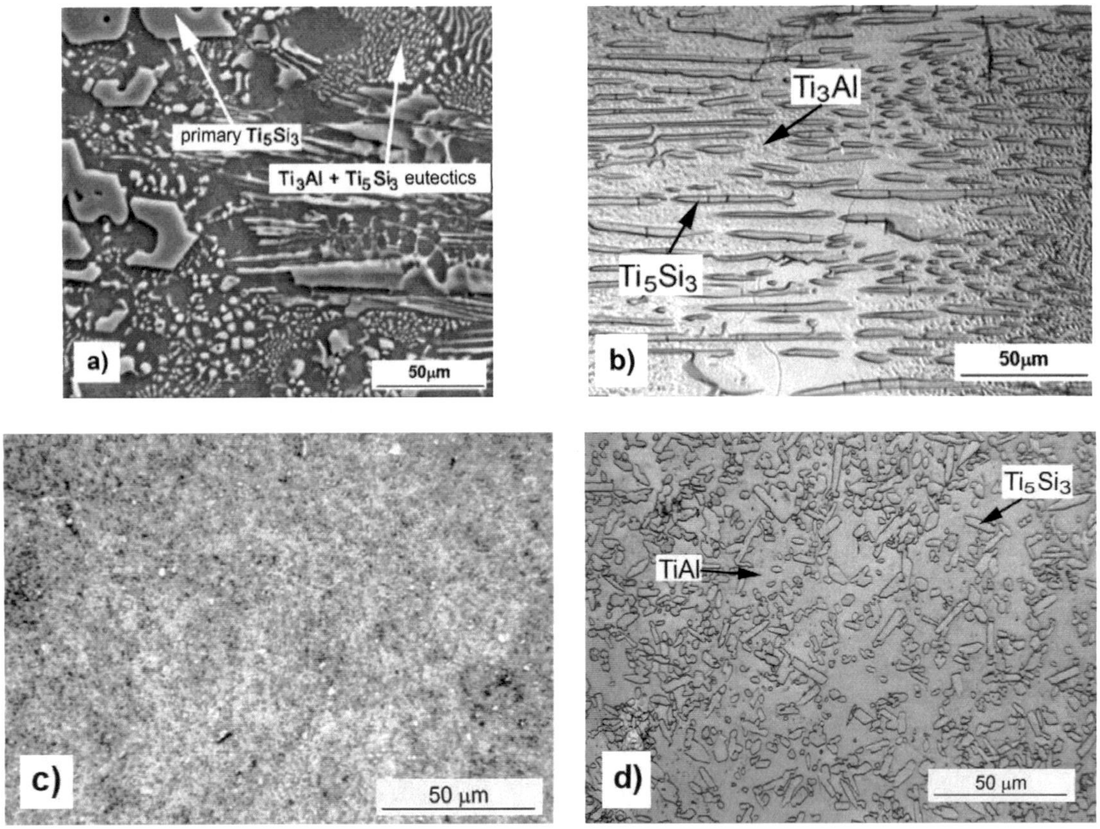

Figure 4: Microstructure of the Ti-Al-Si alloy prepared by: a) casting, b) rapid solidification, c) mechanical alloying and spark plasma sintering, d) reactive sintering [10,17]

After optimization of the conditions, the porosity of less than 7 and 4 vol. % was achieved for Ti-Al-Si and Fe-Al-Si alloys, respectively [10,11] (Figure 4d). Further lowering of the porosity can be achieved by use of pressure-assisted reactive sintering. In addition, the products of this technology are very fine in comparison with the common cast materials. The particles of silicides and ternary phases reach rounded shapes on the contrary to cast alloys (Figure 4a). It positively influences the mechanical and tribological properties [17].

5. Conclusion

Intermetallics are already used as substitutes of CRM-containing materials in selected applications (e.g. jet engines and furnace elements) and their importance will probably grow. However, the extensive development and positive reaction of the industry is still needed. For this purpose, the CA15102 action is a very important tool. The modern powder metallurgy technologies were presented as efficient tools for processing of technically important intermetallics.

Acknowledgement
International frame of the research was supported by COST Action CA15102. This research was also financially supported by Czech Science Foundation, project No. 17-07559S.

References

[1] Critical Raw Materials, https://ec.europa.eu/growth/sectors/raw-materials/specific-interest/critical_fr (accessed 29.10.2017).
[2] MASSALSKI, T.B. (1990) *Binary Alloy Phase Diagrams*, ASM International, Materials Park, USA.
[3] BEWLAY, B. P., NAG, S., SUZUKI, A., WEIMER, M. J. (2016) TiAl alloys in commercial aircraft engines. In: *Materials at High Temperatures*, Vol. 33, pp. 549-559.
[4] NOVÁK, P., KŘÍŽ, J., MICHALCOVÁ, A., VOJTĚCH, D. (2014) Microstructure Evolution of Fe-Al-Si and Ti-Al-Si Alloys during High-Temperature Oxidation. In: *Powder Metallurgy*, Vol. 782, pp. 353-358.
[5] NOVÁK, P., ŠERÁK, J., VOJTĚCH, D., ZELINKOVÁ, M., MEJZLÍKOVÁ, L., MICHALCOVÁ, A. (2011) Effect of alloying elements on microstructure and properties of Fe-Al and Fe-Al-Si alloys produced by reactive sintering. In: *Key Engineering Materials*, Vol. 465, pp. 407-410.
[6] GARCÍA-ALONSO, M.C., LÓPEZ, M.F., ESCUDERO, M.L., GONZÁLEZ-CARRASCO, J.L., MORRIS, D.G. (1999) Corrosion behaviour of an Fe_3Al-type intermetallic in a chloride containing solution. In: Intermetallics, Vol. 7, pp. 185-191.
[7] NOVÁK, P., ŠERÁK, J., VOJTĚCH, D., ZELINKOVÁ, M., MEJZLÍKOVÁ, L., MICHALCOVÁ, A. (2011) Effect of alloying elements on microstructure and properties of Fe-Al and Fe-Al-Si alloys produced by reactive sintering. In: *Key Engineering Materials*, Vol. 465, pp. 407-410.
[8] NOVÁK, P., ŠOTKA, D., NOVÁK, M., MICHALCOVÁ, A., ŠERÁK, J., VOJTĚCH, D. (2011) Production of NiAl–matrix composites by reactive sintering. In: *Powder Metallurgy*, Vol. 54, pp. 308-313.
[9] SCHINDLER, I., KRATOCHVÍL, P., PROKOPČÁKOVÁ, P., KOZELSKÝ, P. (2010) Forming of cast Fe – 45 at.% Al alloy with high content of carbon. In: Intermetallics, Vol. 18, pp. 745–747
[10] VOJTĚCH, D., NOVÁK, M., NOVÁK, P., LEJČEK, P., KOPEČEK, J. (2008) Unidirectional crystallization and high-temperature oxidation of in-situ $Ti_3(Al,Si)$-$Ti_5(Si,Al)_3$ composite. In: *Materials Science and Engineering A*, Vol. 489 pp. 1-10.
[11] NOVÁK, P., VODĚROVÁ, M., HENDRYCH, R., KUBATÍK, T., MICHALCOVÁ, A., VOJTĚCH, D. (2013) Preparation of aluminium-based quasicrystals. In: *Manufacturing*

Technology, Vol. 13, pp. 390-394.

[12] SKIBA, T., HAUŠILD, P., KARLÍK, M., VANMEENSEL, K., VLEUGELS, J. (2010) Mechanical properties of spark plasma sintered FeAl intermetallics. In: *Intermetallics*, Vol. 18, 1410-1414.

[13] KNAISLOVÁ, A., NOVÁK, P., CYGAN, S., JAWORSKA, L., CABIBBO, M. (2017) High-Pressure Spark Plasma Sintering (HP SPS): A Promising and Reliable Method for Preparing Ti–Al–Si Alloys. In: *Materials*, Vol. 10, 465.

[14] NOVÁK P., MICHALCOVÁ A., ŠERÁK J., VOJTĚCH D., FABIÁN T., RANDÁKOVÁ S., PRŮŠA F., KNOTEK V., NOVÁK M. (2009) Preparation of Ti-Al-Si alloys by reactive sintering. In: *Journal of Alloys and Compounds*, Vol. 470, pp. 123–126.

[15] NOVÁK, P., KNOTEK, V., ŠERÁK, J., MICHALCOVÁ, A., VOJTĚCH, D.: (2011) Synthesis of Fe–Al–Si intermediary phases by reactive sintering. In: *Powder Metallurgy*, Vol. 54, pp. 167-171.

[16] NOVÁK, P., MICHALCOVÁ, A., MAREK, I., VODĚROVÁ, M., VOJTĚCH, D. (2012) Possibilities of the observation of chemical reactions during the preparation of intermetallics by reactive sintering. In: *Manufacturing Technology*, Vol.12, pp. 197-201.

[17] NOVÁK, P., KNOTEK, V., VODĚROVÁ, M., KUBÁSEK, J., ŠERÁK, J., MICHALCOVÁ, A., VOJTĚCH, D. (2010) Intermediary phases formation in Fe–Al–Si alloys during reactive sintering. In: *Journal of Alloys and Compounds*, Vol. 497, pp. 90-94.

IOP Conf. Series: Materials Science and Engineering **329** (2018) 012014 doi:10.1088/1757-899X/329/1/012014

Application of Electro Chemical Machining for materials used in extreme conditions

Z Pandilov

Faculty of Mechanical Engineering, University "Ss. Cyril and Methodius"-Skopje, Republic of MACEDONIA

E-mail: panzo@mf.edu.mk

Abstract. Electro-Chemical Machining (ECM) is the generic term for a variety of electro-chemical processes. ECM is used to machine work pieces from metal and metal alloys irrespective of their hardness, strength or thermal properties, through the anodic dissolution, in aerospace, automotive, construction, medical equipment, micro-systems and power supply industries. The Electro Chemical Machining is extremely suitable for machining of materials used in extreme conditions. General overview of the Electro-Chemical Machining and its application for different materials used in extreme conditions is presented.

1. Introduction

Electro Chemical Machining, abbreviated ECM, is an advanced metal-working technique which can machine products that are difficult or impossible to design through conventional machining [1,2]. It is an extremely accurate technique, capable of machining any electrically conductive work piece due to the fact that the technology is based on electrolysis (i.e., chemical change, especially decomposition, produced in an electrolyte by an electric current). ECM is used in aerospace, automotive, construction, medical equipment, micro-systems and power supply industries.

Almost all kinds of metal can be electro-chemically machined, especially high-alloyed nickel - or titanium-based ones, as well as hardened materials [3].

As it is a contactless procedure with no heat input, the process is not subject to any of the disadvantages experienced with traditional machining methods, e.g. tool wear, mechanical stresses, micro-fissures caused by heat transfer or the need for subsequent deburring operations. All electro-chemical machining processes are characterized by stress-free stock removal, gentle transitions and top-quality surfaces without burr formation.

During the process the metal work piece is dissolved (Machining) locally through electricity (Electro) and chemistry (Chemical) until it reaches the required complex 3D end shape.

2. Principle of Electro Chemical Machining

Electro Chemical Machining is based on the principle of electrolysis. The ECM tool (cathode) is positioned close to the work piece (anode) and a low-voltage, high-amperage direct current is passed between them via an electrolyte flowing through the anode-cathode gap (figure 1) [3]. Material is removed by anodic dissolution and is carried away by the electrolyte. A charge exchange takes place between the cathode and the anode in an aqueous electrolyte solution which targets specific areas of the workpiece. This can be used to create contours, ring ducts, grooves or bell hollows with no contact, but with very high precision. The removed material is precipitated from the electrolyte solution in the form of metal hydroxide. The machining can be carried out regardless of the structural condition of the metal and both soft and hard materials, including exotic metals, can be machined.

The components are subjected to neither thermal nor mechanical stresses.

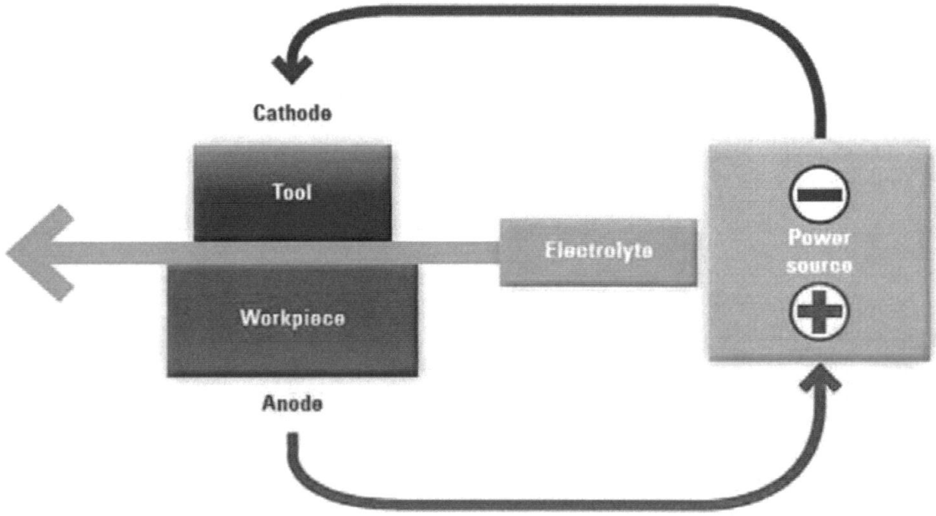

Figure 1. Functional principle of Electro-Chemical Machining [3]

The principle scheme of Electro-Chemical Machining process is presented in figure 2.

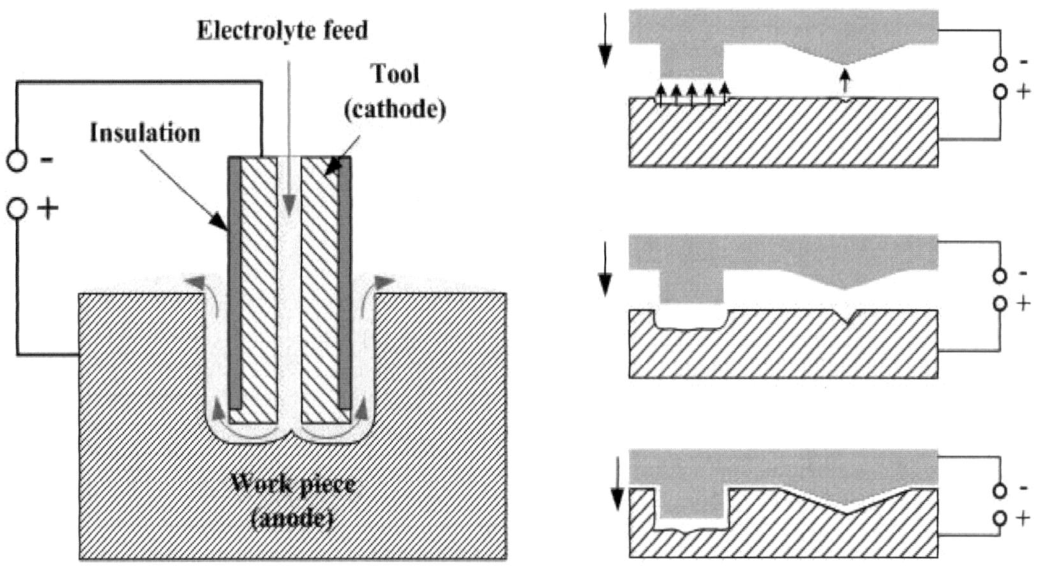

Figure 2. The principle scheme of Electro-Chemical Machining process [4,5]

The work piece is mounted in a fixture electrically isolated from the tank and other machine parts. The work piece is connected to the positive terminal (anode) of the Power Supply. The tool is connected to the negative terminal (cathode).The electrolyte is continuously flowing through a hole in the tool to the gap between the work piece and the tool surfaces. The tool is moving towards the work piece at a constant speed. The gap between the tool and the work piece is kept constant. Stable behavior of the process is a result of a control of the power supply voltage.The final shape of the work piece formed as a result of the electrochemical machining process conforms the shape of the tool.

3. Electrolyte in the Electro Chemical Machining

The electrolyte in the processes of Electro Chemical Machining has several functions: the current is carried between the tool and the workpiece through electrolyte, the produced heat is dissipated by the liquid electrolytic solution, the product of machining is removed by the solution and it keeps the reactions continuous by supplying the elements necessary for the reaction.

For different types of machined materials different types of electrolytes are used in Electro Chemical Machining:

-Sodium chloride (NaCl) at the concentration of 20% - for ferrous alloys (e.g. Steels and cast irons and cobalt alloys).

-Sodium nitrate ($NaNO_3$) - for ferrous alloys.

-Hydrochloric acid (HCl) - for nickel alloys.

-A mixture of sodium chloride (NaCl) and sulfuric acid (H_2SO_4) - for nickel alloys.

-A mixture of 10% hydrofluoric acid (HF), 10% hydrochloric acid (HCl), 10% nitric acid (HNO_3) - for titanium alloys.

-Sodium hydroxide (NaOH) - for tungsten carbide (WC).

The selection of the electrolyte should be done by considering the following matters: required machining rate, required dimensional accuracy and surface texture and integrity.

The electrolyte should possess several important properties [6]:

-Electrolyte must posses high electrical conductivity.
-Electrolyte should posses high specific heat.
-The viscosity must be as low as possible.
-The electrolyte must be chemically stable.
-Electrolyte must be chemically active to cause the better metal or material removal rate.
-Electrolyte should not form any type of excess layer on the top of electrolyte, tool or the work piece.
-Electrolyte should not be toxic and corrosive.
-Electrolyte should be economical and easily available.

Perfect electrolyte flow across the machining tool is mandatory for proper machining. Cavitation is likely to be occurred in the tool. So proper care is necessary to keep the tool in shape. Tool design must ensure the uniform flow of electrolytic solution in all machining areas. Optimum flow of the electrolyte is desired because excessive flow can cause erosion of the tool.

Mainly two types of flows are used: divergent flow and convergent flow.

Convergent flow provides a smoother flow of electrolytes.

The advantages of convergent flow system are:

-improved surface finish
-improved uniform and predictable side over cut as well as front machining gap,
-less prone to arcing,
-clean operating environment and
-stray currents make it possible to be eliminated unwanted machining.

But it is also important to be mentioned that machining in convergent flow is much expensive than divergent.

4. Components of Electro Chemical Machining equipment

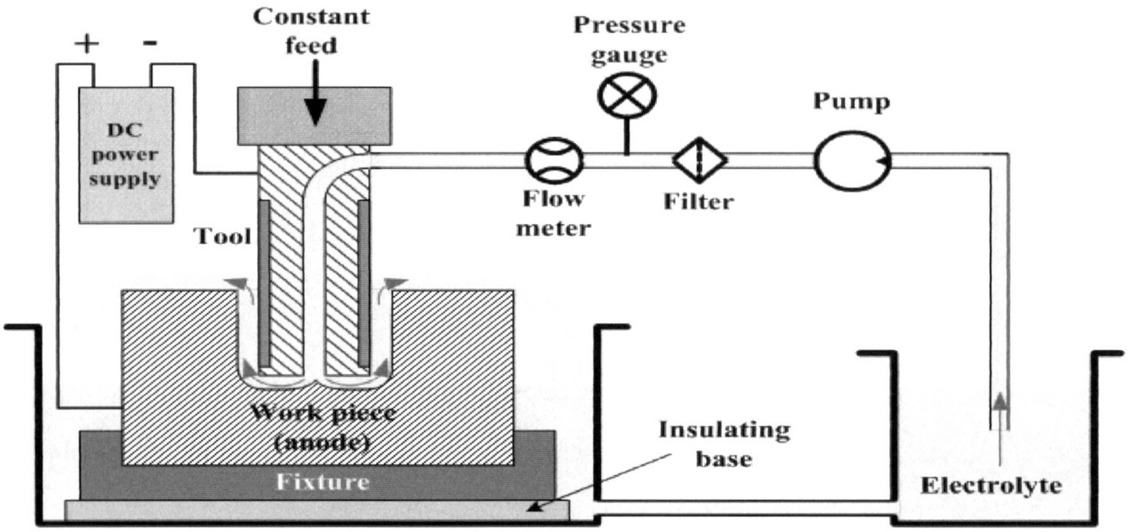

Figure 3. The components of Electro Chemical Machining equipment [4,5,6]

An industrial Electro Chemical Machining equipment consists of the following 4 systems (components) (figure 3) [4,5,6]:

- DC Power supply. The machining rate in electrochemical machining is proportional to the electric current density. In order to achieve high values of the machining rate electrochemical machining is commonly performed at a high direct current exceeding 1000 A. The gap between the tool and the work piece must be low for high-pitched correctness, thus the voltage must be small to prevent a short circuit. The voltage of the process is 5-25 V. The control system uses some of this electrical power.
- Electrolyte circulation system. The products of the electrochemical reaction should be removed from the gap between the work piece and the tool. Accumulation of the reaction products causes decrease of the process efficiency and reduction of the rate of machining. Therefore the electrolyte flow speed should be high. Commonly it is in the range 300-3,000 m/min (1,000-10,000 ft/min). The inlet pressure should be in between 0.15-3 MPa. The electrolyte system should comprise a fairly strong pump. The electrolyte is continuously filtered in order to trap the precipitated reaction products (sludge).
- Control system. Electrical parameters of the process (voltage and current)), tool feed speed and parameters of electrolyte circulation system (inlet and outlet pressure of electrolyte, temperature of electrolyte) are controlled by the control system, which provide stable and efficient operation of the unit
- Mechanical system. It consist of the table, the frame, work enclosure (prevents the electrolyte from spilling), the work head (where the tool is mounted).The tools (electrodes) are also part of the mechanical system. One of the most important parameters of electrochemical machining is maintaining a constant voltage level. This is achieved by the control system providing a movement of the tool at a constant speed equal to the linear rate of machining. The process in a steady state is performed at a constant (typically gap 0.1-0.4 mm / 0.004-0.016"). A firm fixation of the work piece provided by the fixture, the table and the frame is also important for stable operation of the system at a constant gap. Conventional machining equipment including CNC machines may be modified for electrochemical machining process.

5. Applications of Electro Chemical Machining

Electro Chemical Machining is mainly used in the areas where conventional machining techniques are not feasible [7]. Theoretically Electro Chemical Machining could be used for machining the all electro conductive metals and alloys. The usual applications of Electro Chemical Machining are:

- Machining of hard materials and heat resistant materials. The process parameters and the tool life time do not depend on the hardness of the work piece. Therefore electrochemical machining is often used for machining hard materials.
- Machining of cavities in forging dies, drilling deeper holes and irregular shaped holes which can not be obtained by conventional machining methods.
- Machining of complex profiles like turbine wheels, turbine and jet blades.
- Die sinking. Electro Chemical Machining is often used as an alternative to the cavity type electric discharge machining (EDM).
- Fabrication of thin walled parts. Electrochemical machining does not produce surface stress in the work piece therefore even very brittle and easily deformed materials may be machined in thin walled shapes.
- Grinding of a work piece by a rotating wheel, which performs grinding operation through an electrolyte. The wheel is conductive and catholically connected. Non-conductive hard particles are set on the wheel surface. The particles provide a constant gap through which an electrolyte is continuously fed. Hard and brittle materials are ground by the method.
- Rough corners or edges can be turned into very smooth parts and the process is known as deburring.

6. Materials possible to be processed with Electro Chemical Machining (materials used in extreme conditions)

With Electro Chemical Machining is possible processing of practically all groups of constructional steels and alloys used in the industry, including high-strength steels and alloys, metal-ceramics, nanostructure alloys.

This materials could be easily machined by Electro Chemical Machining [8]: alloy steels and construction steels, corrosive resistant steels, high-carbon chromium steels, stainless chrome-nickel steels, tool steels, chromium-nickel heat-resistant alloys, HSS, copper and copper alloys, nickel and nickel alloys, magnets, titanium and its alloys, ceramic-metal hard alloys (WC-Co, WC-TiC-Co), nanostructure steels and alloys.

Most of these materials with exception of copper, copper alloys and magnets could be used in extreme conditions of temperature, loading, friction, wear, corrosion, in Energy, Transportation and Machinery manufacturing industries.

7. Conclusion

Electro Chemical Machining has a significant place among the methods of machining of conducting materials. It is applied in cases where alternative methods (especially the traditional ones) encounter problems: for operations which are time and hand-labour consuming. Electro Chemical Machining is feasible for the machining of difficult-to-cut materials (high-strength, tough, hard, and brittle alloys, super-alloys), complex-shaped parts, and the machining in the difficult-to-reach areas. In some cases, Electro Chemical Machining enables one to perform unique operations, for example, drilling of very small, deep, curved holes. The Electro Chemical Machining processes are in progress, spread into new areas of modern technologies, for instance, micromachining.

Electro Chemical Machining today is widely used in processing of materials used in extreme conditions of temperature, loading, friction, wear, corrosion, in Energy, Transportation and Machinery manufacturing industries.

References

[1] *Encyclopedia of Electrochemistry* 2007 (Editors: Allen J Bard & Martin Stratmann) Volume **5** Electrochemical Engineering Wiley-VCH Weinheim ISBN 978-3-527-30610-7.

[2] Wüthrich R and Fascio V 2005 Machining of non-conducting materials using electrochemical discharge phenomenon—an overview *International Journal of Machine Tools and Manufacture* Volume **45** Issue 9 pp 1095-1108

[3] Deconinck D, Damme S Van and Deconinck J 2012 A temperature dependent multi-ion model for time accurate numerical simulation of the electrochemical machining process Part I: Theoretical basis *Electrochimica Acta* Volume **60** 15 pp 321-328

[4] Kozak J, Rajurkar P K and Makkar Y 2004 Study of Pulse Electrochemical Micromachining *Journal of Manufacturing Processes* Volume **6** Issue 1 pp 7-14

[5] Chaudhary D and Nara J 2014 Electrochemical Machining *Journal of Emerging Technologies and Innovative Research (JETIR)* (Volume **1** Issue 7) pp 620-624 (ISSN-2349-5162)

[6] Kalpakjian S and Schmid R S 2013 *Manufacturing Engineering and Technology* 7th edition (Singapore: Pearson Publications) ISBN: 9780133128741

[7] Rajurkar P K, Sundaram M M and Malshe P A 2013 Review of Electrochemical and Electrodischarge Machining *Procedia CIRP* Volume **6** pp 13-26

[8] El-Hofy H 2005 *Advanced machining processes: nontraditional and hybrid machining processes* McGraw Hill Professional

IOP Conf. Series: Materials Science and Engineering **329** (2018) 012015 doi:10.1088/1757-899X/329/1/012015

Fracture behaviour of WC-Co hardmetals with WC partially substituted by titanium carbide

M. Szutkowska[1], M. Boniecki[2], S. Cygan[1], A. Kalinka[1], M. L. Grilli[3], S. Balos[4]

[1] Institute of Advanced Manufacturing Technology, Cracow, Poland
[2] Institute of Electronic Materials Technology, Warsaw, Poland
[3] ENEA Energy Technology Department, Rome, Italy
[4] Faculty of Technical Sciences, Novi Sad, Serbia
E-mail: szutkows@ios.krakow.pl

Abstract

The addition of various amounts of $TiC_{0.9}$ phase in the range from 5wt.% to 20wt.% substituting WC phase was applied in WC-Co hardmetals with 9.5 wt.% bonding cobalt phase. The hardmetals were consolidated using Hot Isostatic Pressing (HIP) method at temperature of 1573K and pressure of 1500 atm. The plain strain fracture toughness has been determined from 3PB test on a pre-cracking single edge notched beam (SENB) specimen. The indentation fracture toughness with Vickers cracks for comparison was also measured, which changed from 12 to 9.0 $MPa{\cdot}m^{1/2}$. The amount of the $TiC_{0.9}$ phase affected the mechanical and physical properties: Vickers hardness from 12.5 to 14.0 GPa, Young's modulus from 550 to 460 GPa, density from 13.1 to 9.6 g/cm^3, friction coefficient from 0.24 to 0.45, fracture toughness from 16.8 to 11.0 $MPa{\cdot}m^{1/2}$. Scanning electron microscopy (SEM), X-ray and electron diffraction phase analysis were used to examine the WC-Co hardmetal with addition of the $TiC_{0.9}$ phase. For comparison, physical and mechanical properties of the WC-Co hardmetals before modification were tested.

Keywords: Hardmetals, sintering HIP, fracture toughness, Palmqvist cracks, titanium carbide

1. INTRODUCTION

Hardmetals or cemented carbides (sometimes less accurately called sintered carbides) are among the most commonly and most frequently used tool materials produced by powder metallurgy. The hardmetals are used to designate a metal matrix composite constituted by hard ceramic particles, normally WC, into the metallic matrix [1]. Now more than 90% of all WC-hardmetals utilize Co as the preferred binder metal with contents between 6 and 25 wt.%. The apparent superiority of cobalt, relatively to other binders, is related to its best comminution characteristics in milling, superior wettability for WC, higher solubility of WC in cobalt at sintering temperatures and excellent properties [2]. Cobalt is a ferromagnetic metal with a hcp structure as the most stable phase at room temperature. It undergoes a phase transition from hcp into a fcc structure at 450 °C. The reasons for the dominant role of Co are also some unique properties of Co and the W–C–Co ternary system [3]. As a hard ceramic phase in WC-Co hardmetals are commonly used tungsten carbides (WC) because of their readily wettability for Co, ultimate compression strength of about 2.7 GPa, high melting point at 2870°C, extremely large Young's modulus of approximately 530-700 GPa and very good thermal conductivity of 110 $W \cdot m^{-1}K^{-1}$. Tungsten carbide is very hard, ranking about 9 on Mohs scale, and with a Vickers hardness of around 26.0 GPa [4]. WC and Co form a pseudo-binary eutectic in a three component system (W, C and Co). The phase diagram also shows that there are so-called η-carbides with composition $(W,Co)_6C$ that can be formed and the brittleness of these phases makes control of the carbon content in WC-Co cemented carbides important [5]. If the processing conditions are such that the carbon balance is either too high or too low then other phases are present in the structure, *i.e.* graphite if the C is high and eta-phase (an M_6C or $M_{12}C$, where M is generally Co_nW_n) if the C content is low. The structure of two phases WC-Co hardmetals is defined by three key parameters: grain size of the WC phase, volume fraction and composition of the Co binder phase [6]. In WC-Co hardmetals the tungsten carbides are commonly used in quantities ranging from 75 to 94 wt.%. Such a characteristic structure for hardmetals, with the presence of a ductile bonding phase and a hard and brittle carbide phase, gives combination in one material of contrasting features like: high wear resistance and hardness with high strength and good ductility. The hardmetals have the potential to change its composition so that the resulting physical and chemical properties ensure maximum resistance to wear, deformation, fracture, corrosion, and oxidation. Transverse rupture strength improves with increasing cobalt content up to a maximum of 20±5 wt.% Co. The maximum strength depends strongly on the variables, such as carbide grain size. The most important properties of the hardmetals used for mining tools are wear resistance and toughness. These properties strongly depend on content of the Co binder phase (Fig.1) [7].

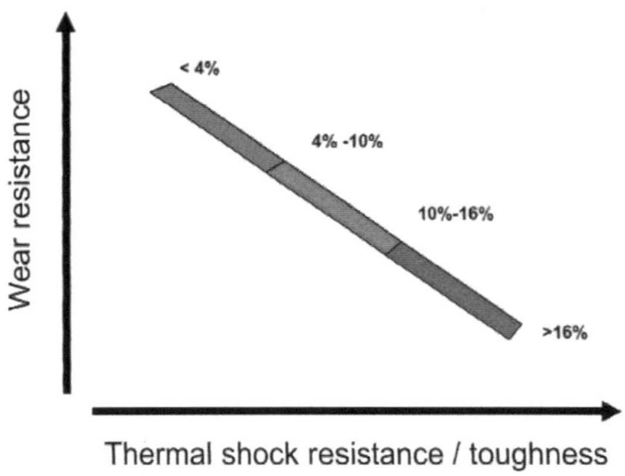

Fig.1. Wear resistance as a function of thermal shock resistance/toughness of the WC-Co hardmetals with various content of cobalt.

During the wear or fatigue processes hardmetals produce fracture from crack-like defects that develop in service. The plane strain fracture toughness of the hardmetals is represented by its critical stress intensity factor K_{IC}. Consequently, fracture toughness is an important parameter that influences the residual strength of hardmetal components. The fracture toughness depends on content of cobalt and this property increases with an increase in the cobalt content, and with increased WC grain size (Fig.2) [7].

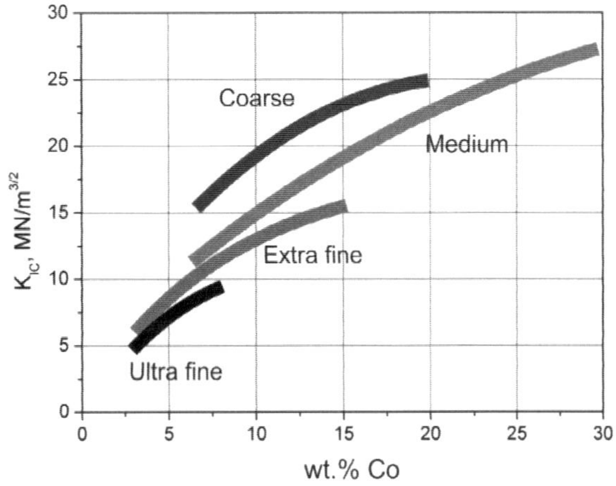

Fig.2. Plain strain fracture toughness (K_{IC}) as a function of cobalt content for different grain sizes.

Measurement of the plain strain fracture toughness for the hardmetals is difficult because there is no agreed standard test method for obtaining accurate values of K_{IC}. This is particularly important for materials with K_{IC} values greater than about 15 MN m$^{-3/2}$ that are difficult to pre-crack. Typical values of plane strain fracture toughness lie in the range 7 - 25 MN m$^{-3/2}$. These values are much higher than for similar hard materials like oxide ceramics which have fracture toughness of 3 - 7 MN m$^{-3/2}$ [6, 8]. ISO Standard [9] gives possibility to determine the Palmqvist toughness by so called the indentation fracture (IF) tests which is widely used to assess the toughness of hardmetals only for small amounts of material and allows to omit the difficulties of pre-cracking conventional toughness test-pieces. But due to particularly sensitivity to the test-piece preparation, the procedure requires re-examination for very fine grained hardmetals. A method commonly used for determining the plane strain fracture toughness of the hardmetals bases on pre-cracking conventional toughness test-pieces with notch prepared in different way (wedge indentation and fatigue SEPB specimen, beam with notch-SENB, Single Edge V-notched Beam-beam containing notch with sharpened tip diamond honed, Chevron Notched Beam-CNB and Chevron Notched Short Rod–CNSR) in the three point bending test (3PB) [6]. Unusual combinations of otherwise quite different components led to permanent technical acceptance of these materials since their discovery. Currently, tungsten carbide products are used in various industrial sectors; nearly 67% of the total hardmetals production is dedicated to cutting tools, about 13% for mining, oil extraction and tunneling, and 11% and 9% for wood and construction respectively [10]. But the WC-Co hardmetals contains tungsten carbide (WC) and cobalt (Co), both of which are defined by the EU as critical and, in the case of cobalt, dangerous to health [11,12]. The EU critical materials list contains those substances of significance to the EU economy but whose availability is at great risk, and are mostly non-renewable. Due to the significant growth of price, particularly in the recent years, of WC-Co hardmetals blends, the chemical composition of WC-Co hardmetals has been changed to reduce the WC content and substitute it with other ceramic components. It is also significant that 85%

of the world's total tungsten production is located in China, and this undoubtedly has an impact on the continuous increase in product prices [13,14]. Because titanium has a much lower risk of supply than tungsten in terms of its reserves and distribution, cubic carbides as TiC partially substituted for the WC is a good solution. Titanium carbide, with a very high melting temperature of 3065°C, has excellent properties such as: high Vickers hardness 28-32 GPa, Young's modulus of 448-451 GPa, low density of 4.94 g/cm^3, good high-temperature strength, and good corrosion resistance [15]. Additionally, no detrimental new phase is formed except that W diffuses into the TiC lattice to form a (Ti,W)C solid solution according to the TiC-W binary alloy diagram [16]. Partial replacement of WC by cubic TiC improved the high temperature strength, however, no effect on the mechanical properties was investigated [17]. Modification of WC-Co hardmetals towards the partial substitution of tungsten carbide (WC) by addition of TiC phase was the main goal of this work. This requires the change in technology of new materials by varying the pressure and temperature of sintering.

2. Experimental Procedure

A starting material was commercially available. Mixture of WC-Co powders signed as YK15.6 was delivered by the Chinese producer Chongy and Zhangyuan Tungsten Co. Ltd. According to the certificate of quality the WC-Co mixture contained 9.5 wt.% of cobalt, density of 14.57 g/cm^3 and Rockwell hardness 87.6 HRA. The grain size of the WC-Co mixture (YK15.6) was measured with the Shimadzu SA-CP3 type equipment. Density of powder was determined using AccPyc 1340 helium pycnometer, manufactured by Micrometrics Inc. The powder mixture, of 14,59 g/cm^3 density was 3.74 μm in average grain size (median), 3.53 μm in modal value and 0.730 m^2/g in surface development. The $TiC_{0.9}$ powder, of 4.93 g/cm^3 density, was 4.46 μm in average grain size (median), 15.0 μm in modal value and 1.763 m^2/g in surface development. This powder was processed by the relatively low-cost Self Propagating High Temperature Synthesis (SHS) method at the Faculty of Materials Science and Ceramics of the AGH University of Science and Technology. SEM micrographs of tested powders are presented in Fig. 3.

a) b)

Fig. 3. SEM micrographs of tested powders: a) WC-Co; b) $TiC_{0.9}$

Mixture of the WC-Co powders with addition of the $TiC_{0.9}$ phase in the range from 5wt.% to 20wt.% substituting WC phase was prepared in planetary ball mill of Pulverisette 6 mono, produced by Fritsch, through 2 hours with presence of acetone as lubrication medium. 250 ml

WC milling bowl with of 5 mm in diameter WC millers was used. Green bodies were consolidated using Hot Isostatic Pressing (HIP) method at the temperature of 1573K and pressure of 152 MPa in GONAR Company. Specimens with size of 5.0x6.5x20 mm were used for various tests. After sintering specimens were lapped on a cast iron disk, ground with DIA-Pro type diamond abrasive (size of diamond grains about 9 μm) and polished with a OP-S type diamond abrasive (size of diamond grains about 0.04 μm). Following physical and mechanical properties were measured: density ρ, Young's modulus E, friction coefficient μ, Vickers hardness HV, plane strain fracture toughness K_{IC} and indentation fracture toughness $K_{IC(HV)}$. The density was determined hydrostatically, by weighing the specimen in air and in distilled water according to PN-ENISO standard [18]. The hardness was measured by the Vickers method under the load of 294.3 N, using a Future Tech FLC-50VX Vickers hardness tester. For each sample, five indentations were done. Young's modulus of the composites was determined with using a Panametrics Epoch III ultrasonic flaw detector. The measurements were based on the transition velocity of the ultrasonic waves through the sample. The velocities of transversal and longitudinal waves were designated as the ratio of the sample thickness and the relevant transition time. Because, the error in the thickness measurements was ±0.01 mm and in the time-of-flight measurements (±1 ns), the resulting error in the ultrasonic velocity was about 1%. Constantly, the error of Young's modulus measurement has obtained the value of 2%. Calculations were made using the formula:

$$E = \rho C_T^2 \frac{3C_L^2 - 4C_T^2}{C_L^2 - C_T^2} \qquad (1)$$

where: C_L — velocity of the longitudinal wave and C_T — velocity of the transversal wave, ρ- density of the material.

Plane strain fracture toughness, K_{IC} was a value obtained from tests on a specimen with a well-defined geometry of crack. Fracture toughness was assessed by means of the conventional method (3PB) on pre-cracked SENB (Single Edge Notched Beam) specimens with 1.5 mm × 4.0 mm × 15.0 ± 0.1 mm dimensions An initial notch 0.9 mm deep was produced by a diamond saw (thickness 0.20 mm) and then the notch tip was pre-cracked with a thin diamond saw(thickness 0.025mm). The total initial notch length was approximately 1.1 mm. The relationship K_{IC}= f(c) is given by the Eqs. (2, 3).

$$K_{IC} = 1.5 \frac{P_c S}{W^2 B} Y c^{1/2} \qquad (2)$$

$$Y = \frac{\sqrt{\Pi}}{(1-\beta)^{\frac{3}{2}}} \left[0.3738\beta + (1-\beta) \sum_{i,j=0}^{4} A_{ij} \beta^i \left(\frac{W}{S} \right)^j \right] \qquad (3)$$

where: P_c—critical load, S—support span, W—width and B—specimen thickness, Y — geometric function, c — crack length, β — c/W and A_{ij} is the coefficients given by Fett [19].

Both the Palmqvist and the median/radial cracks propagating from the corner of the Vickers hardness indentation provide base to determination the indentation fracture toughness $K_{IC(HV)}$.

Relation between the fracture toughness and the ratio of cracks to indent size at the condition $c/a \geq 2.5$ is expressed by the equation (4) [20,21]:

$$(K_{IC}\ \varphi\ /H\ a^{1/2})\ (H/E\ \varphi)^{2/5} = 0.129\ (c/a)^{-3/2} \quad (4)$$

where: K_{IC} is the critical stress intensity factor, φ is the constrain factor, H is the Vickers hardness, E is the Young's modulus, a-half of indent diagonal.

The right side of the equation (4) is equal to $0.035\ (c/a)^{-1/2}$ for the Palmqvist cracks (Fig.4).

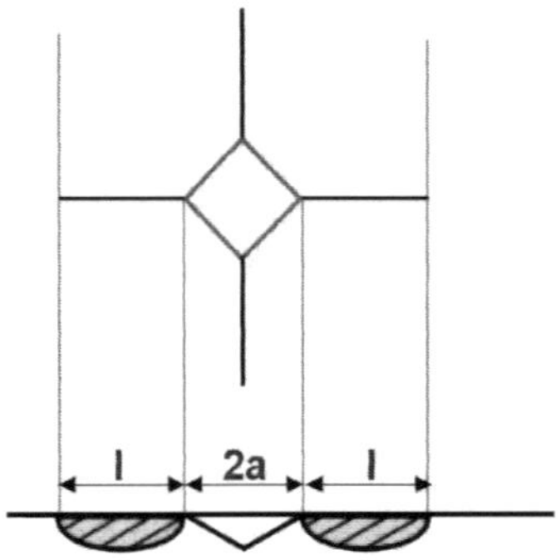

Fig. 4. Palmqvist cracks formed during an indentation toughness test.

The friction coefficients, in sliding contact with a WC ceramic ball (the ball's roughness *Ra* was 0.05 μm), were determined in ball-on-disk tests at loading 25N, using a CETR UMT-2MT universal mechanical tester (USA). The X-ray diffraction patterns were obtained by using a PANalitycal Empyrean diffractometer with copper radiation (λCu Kα = 1.5406 Å). The phase analysis was carried out with using ICDD (PDF-4+ 2016) files. The microstructure investigations were performed using a Jeol JSM-6460LV scanning electron microscope.

3. Results and Discussion

Table 1 shows the results of the change of selected physical and mechanical properties for the WC-Co/$TiC_{0.9}$ composites with various amount of $TiC_{0.9}$. For comparison, values of tested properties for the hardmetal without addition of $TiC_{0.9}$ phase (H0) are also given.

Table.1. Selected physical and mechanical properties of tested WC-Co/$TiC_{0.9}$ hardmetals

Samples	Sample symbol	Vickers hardness HV [GPa]	Young's modulus E [GPa]	Density ρ [g/cm^3]
WC-Co +5wt%TiC	H5	12.7±0.2	549±20	13.09±0.0
WC-Co +10wt%TiC	H10	13.1±0.2	519±20	11.81±0.02
WC-Co +20wt%TiC	H20	13.9±0.2	458±20	9.60±0.03
WC-Co	H0	9.6±0.15	542±20	13.52±0.02

Mechanical and physical properties of tested hardmetals depend on the amount of the $TiC_{0.9}$ phase. The Vickers hardness *HV* increased from 12.7 to 13.9 GPa, Young's modulus *E* decreased from 549 to 458 GPa and density *ρ* from 13.09 to 9.60 g/cm^3 when weight percentage

IOP Conf. Series: Materials Science and Engineering **329** (2018) 012015 doi:10.1088/1757-899X/329/1/012015

of the $TiC_{0.9}$ phase increased from 5 to 20 wt%, respectively. The reduction in the density of WC-Co/ $TiC_{0.9}$ composites resulted with a five times lower $TiC_{0.9}$ density compared to the WC phase. For hardmetals with the highest percentage of $TiC_{0.9}$ (H20) phase substituting the WC phase, the increase in the Vickers hardness in comparison to WC-Co hardmetals (H0) was about 40 %. The most similar values of the Young modulus, density and Poisson ratio to the WC-Co hardmetal were obtained for the hardmetals with the smallest content of $TiC_{0.9}$ (H5). In Table 2 the results of the fracture toughness values for the WC-Co/$TiC_{0.9}$ hardmetals with various amount of $TiC_{0.9}$ and for WC-Co hardmetal without addition of $TiC_{0.9}$ phase (H0) were given.

Table.2. Fracture toughness of the WC-Co/$TiC_{0.9}$ and the WC-Co hardmetals

Specimens	**H5**	**H10**	**H20**	**H0**
Fracture toughness $K_{(IC)}$ [$MPa \cdot m^{1/2}$]	16.8	14.0	11.1	18.5
Indentation fracture toughness $K_{IC(HV)}$ [$MPa \cdot m^{1/2}$]	12.0	10.9	9.0	–

The plane strain fracture toughness, K_{IC} decreased with $TiC_{0.9}$ addition. It was a result of increasing the Vickers hardness of the WC-Co/$TiC_{0.9}$ hardmetals. There was a general inverse trend of the hardness against the fracture toughness.The plane strain fracture toughness of the WC-Co/5%wt.$TiC_{0.9}$ (H5) exhibits the highest value similar to the plane strain fracture toughness of WC-Co hardmetal. The indentation fracture toughness, $K_{IC(HV)}$ shows the same dependence on content of the $TiC_{0.9}$ phase in comparison to the plane strain fracture toughness, K_{IC}. It should be noted that values of the indentation fracture toughness, $K_{IC(HV)}$ are lower about 20-30% than the plane strain fracture toughness, K_{IC} determined on the basis 3PB test. The Palmqvist cracks at the corner of the Vickers indent on the specimen's surface and on the fracture of the WC-Co/$TiC_{0.9}$ hardmetal are presented in Fig.5.

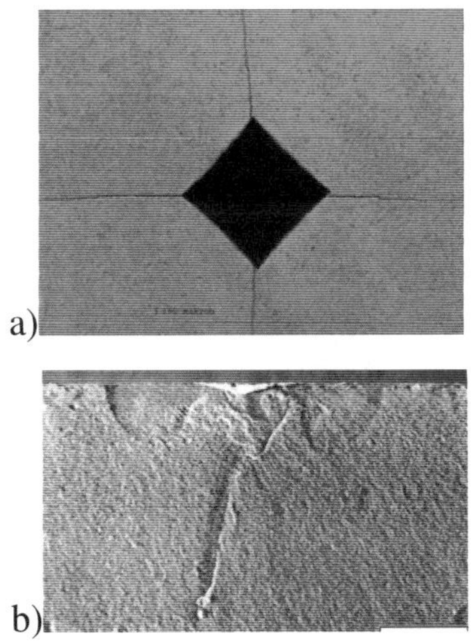

Fig.5. The Palmqvist cracks at the corner of the Vickers indent: a) surface view ; b) fracture view

At low values of Vickers hardness for WC-Co was difficult to determine indentation fracture toughness because of short, deformed cracks which make it difficult to measure the crack length. Friction coefficient of the WC-Co/$TiC_{0.9}$ was within the range from 0.24 to 0.45 for hardmetals with 5%wt. and 20%wt. $TiC_{0.9}$, respectively, Fig. 6.

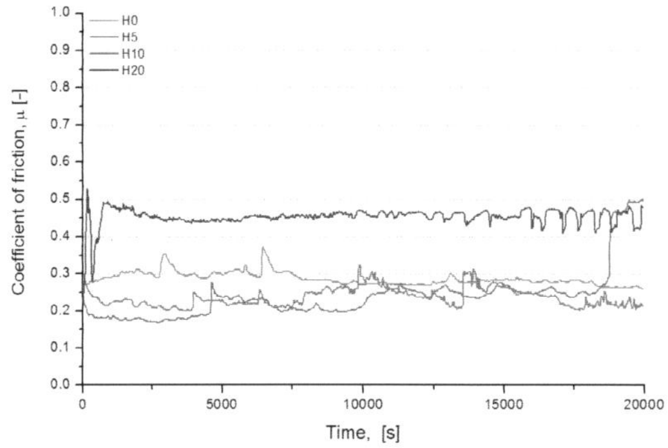

Fig.6. Friction coefficient of the WC-Co/$TiC_{0.9}$ hardmetals with various content of the $TiC_{0.9}$ phase and WC-Co hardmetal (H0).

The SEM microstructure of the WC-Co hardmetal and WC-Co/$TiC_{0.9}$ hardmetals containing an addition of 10% and 20%wt. $TiC_{0.9}$ consolidated using the HIP method is shown in figure 7. The distribution of both types of carbide particles with different shape is homogeneous. Based on the quantitative analysis of elements EDS distributed on the WC-Co/$TiC_{0.9}$ hardmetals with 10 and 20 wt.% (H10, H20) surface, presence of the: α_1WC; α_2WC; β(Ti,W)C, β'(Ti,W)C and γCo phases were observed (Fig.7).

a) b) c)

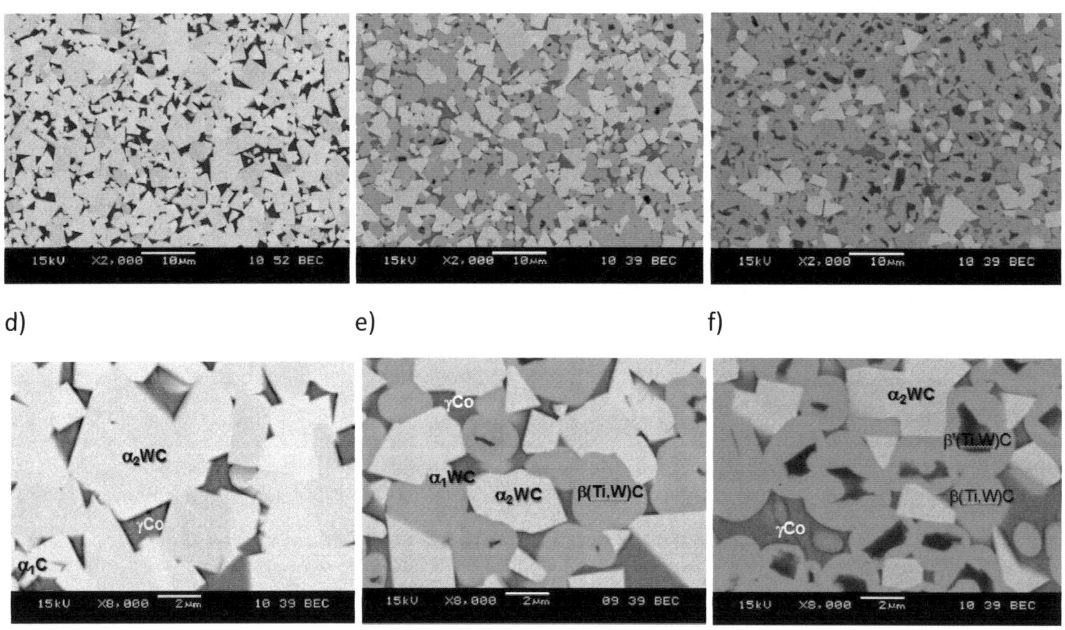

Fig.7. SEM micrograph of tested hardmetals at various magnification 2000 (a,b,c) and 8000 d,e,f): a, d) WC-Co ; b, e) WC-Co/10%wt.$TiC_{0.9}$; c, f) WC-Co/20%wt.$TiC_{0.9}$

SEM microstructure observations of the WC-Co hardmetal revealed the existence of the fine, rounded grains of the α_1WC (secondary solid solutions Co in WC non-recrystallized during sintering), large with geometric shape, regular grains of the α_2WC phase (secondary solid solutions Co in WC recrystallized during sintering) and γCo phase (solid solution of W in Co). Addition of the $TiC_{0.9}$ reveals the new phases in spherical shape, carbides of titanium and tungsten : β(Ti,W)C – carbide of Ti and W enhanced in W and β'(Ti,W)C –carbide of Ti and W depleted in W with higher concentration of Ti. The β'(Ti,W)C cores (dark) are surrounded by a β(Ti,W)C grains (grey). The β(Ti,W)C phases exist in the chain forms. Amount of the β'(Ti,W)C phase increase with increasing of the $TiC_{0.9}$ phase in WC-Co hardmetals. X-ray diffraction of the WC-Co/10%wt.$TiC_{0.9}$ hardmetal which confirms the results of SEM microstructure observations is presented in Fig.8.

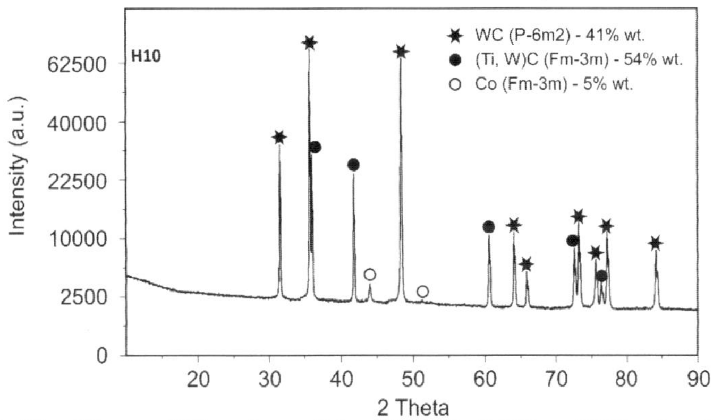

Fig.8. The X-ray diffraction of the WC-Co/10%wt.$TiC_{0.9}$ hardmetal

X-ray diffraction curves obtained from the WC-Co/10%wt.$TiC_{0.9}$ (H10) hardmetal reveal three basic phases: WC of the highest peak intensities due to its highest fraction, then Co and (Ti,W)C with increasing peak intensity when $TiC_{0.9}$ fraction increases from 5 to 20 wt.%. No intermediate phases were observed.

4. Conclusion

- Mechanical and physical properties of tested hardmetals depend on the amount of the $TiC_{0.9}$ phase substituted tungsten carbide in the WC-Co hardmetals.
- Hardmetal with the highest percentage of $TiC_{0.9}$ (H20) phase substituting the WC phase, reveal about 40% increase in the Vickers hardness in comparison to WC-Co hardmetal (H0).
- The inverse trend of the Vickers hardness against fracture toughness is observed.The plain strain fracture toughness of WC-Co/$TiC_{0.9}$ with 5%wt. $TiC_{0.9}$ (H5) exhibits the highest value similar to the fracture toughness of the WC-Co hardmetal.

Acknowledgments

This research was supported by COST Action CA15102 – "Solutions for Critical Raw Materials Under Extreme Conditions" (CRM-EXTREME) and partially by INNOTECH K2/IN2/20/181971/NCBR/13.

References

[1] Ettmayer P 1989 Annu. Rev. Mater. Sci. 19 145-64
[2] Penrice TW 1988 Carbide Tool J. 20/4 12–15
[3] Fernandes C M and Senos A M S 2011 Int. Journal of Refractory Metals and Hard Materials 29 405–418
[4] Groover M P 2010 John Wiley & Sons 135
[5] Ettmayer P and Lengauer P W 1994 John Wiley & Sons
[6] Roebuck B, Gee M, Bennett E G and Morrell R Mechanical Tests for Hardmetals. Measurement Good Practice Guide No. 20. NPL England
[7] Understanding Cemented Carbides. SANDVIK Hard Materials. Destiny Tool. The mark of performance. http://www.destinytool.com/carbide-substrate.html
[8] Szutkowska M 2001 Advances in strength properties testing of sintered tool materials. Proceedings of the European Congress and Exhibition on Powder Metallurgy PM, Nice vol 2 pp 454-460
[9] International Standard IOS 28079:2009 (E) Hardmetals - Palmqvist toughness test.
[10] Houping Z L W 2009 Cemented Carbide 26/2 pp 122-127
[11] Report On Critical Raw Materials For The European Union. Report of the Ad hoc Working Group on defining critical raw materials May 2014. http://www.catalysiscluster.eu/wp/wp-content/uploads/2015/05/2014_Critical-raw-materials-for-the-EU-2014.pdf
[12] Grilli ML, Bellezze T, Gamsjäger E, Rinaldi A, Novak P, Balos S, Piticescu RR, Ruello ML 2017 Solutions for Critical Raw Materials under Extreme Conditions: A Review, Materials, 10, 285.
[13] Silberglitt R, Bartis J T, Chow B G, An D L and Brady K 2013 Critical Materials. Present Danger to U.S. Manufacturing. RAND. National Defense Research Institute. Library of Congress Cataloging-in-Publication Data. ISBN: Copyright 2013 RAND Corporation.
[14] Ishida T, Moriguchi H and Ikegaya A 2011 Sei Technical Review 73 pp 52-55
[15] Bauccio M Ed. 1994 ASM Engineered Materials Reference Book, Second Edition, ASM International, Materials Park, OH
[16] Lipatnikov V N, Rempel A A and Gusev A I 1997 Int. J. Refract. Met. Hard Mater. 15 pp 61-4.
[17] Baratti C, Garcia J, Brito P and Pyzalla A R 2009 Int. J. Refract. Met. Hard Mater. 27 pp 768-776.
[18] PN-EN ISO 3369:2010E Impermeable sintered metal materials and hardmetals - Determination of density.
[19] Fett T 1991 Eng. Fract. Mech. 40/3 pp 683–686.
[20] Niihara K. et al. 1982 Communications of the Amer. Ceram. Soc., C-116
[21] Szutkowska M, Boniecki M 2010 Journal of Optoelectronics and Advanced Materials vol. 12/2 pp 301– 308

IOP Conf. Series: Materials Science and Engineering **329** (2018) 012016 doi:10.1088/1757-899X/329/1/012016

Comparative microstructural and corrosion development of VCrNiCoFeCu equiatomic multicomponent alloy produced by induction melting and spark plasma sintering

É Fazakas [1,7*)], A Heczel [1,2], D Molnár [3,4], B Varga [5], V. Zadorozhnyy [6], Á Vida [1,2,7]

[1] Bay Zoltán Nonprofit Ltd. for Applied Research, Engineering Division, Budapest, Hungary
[2] Institute of Materials Physics, Eötvös University, Budapest H-1117, Pázmány P. stny 1/A
[3] Materials Science Group, Dalarna University, 791 88 Falun, Sweden
[4] Applied Materials Physics, Department of Materials Science and Engineering, Royal Institute of Technology,Stockholm SE-100 44, Sweden
[5] Transilvania University of Brasov, Bulevardul Eroilor 29, Braşov 500036
[6] National University of Science and Technology «MISIS», Leninsky prosp., 4, Moscow, 119049, Russia
[7]Institute for Solid State Physics and Optics, Wigner Research Centre for Physics, H-1525 Budapest, P.O. Box 49, Hungary

Abstract. The present study focuses on the corrosion behavior of a single-phase FCC high entropy alloy (VCrNiCoFeCu) casted by two different methods: induction melting and spark plasma sintering. The corrosion resistance has been evaluated using immersion tests in 3.5% NaCl solution, the potentiodynamic polarization measurements and the results are compared how is dependent the corrosion rate as a function of the production methods. Our results show that induction melted sample is stable in salty environment. On the other hand, based on the changes of polarization curves, there must be an evolution of oxide films on the SPSed sample until reaching the stable oxide layer.

1. Introduction

Traditional concept of alloy systems has been based on utilizing one or two elements as the principal components, and sometimes, minor quantities of other elements are used to enhance the properties, such as steels, NiAl intermetallics and bulk metallic glasses (BMG) [1]. The main reason is that traditional metallurgical knowledge suggests that complex phases or intermetallics will be easily formed in alloy systems with multiple principal components, and thus lead to poor mechanical properties [2].

In the last decade, the so called High-Entropy Alloys (HEAs) became popular materials because of their unique properties, as high strength and ductility simultaneously [1–3], thermal stability [4–6] and excellent corrosion resistant properties [7–9].

The name comes from the fact that high mixing entropy ΔS_{mix} can dominate the term of Gibbs relation: $\Delta G_{mix} = \Delta H_{mix} - T\Delta S_{mix}$. High entropy comes from the high number of elements in the alloy, since based on the definition as an alloy HEA system is composed of five or more kinds of principal elements in an equimolar or near equimolar ratio, with a small difference in atomic radii (<15%) and concentration of each element is varying from 5 to 35 at%. By this way the significance of the enthalpy of mixing (ΔH_{mix}) is oppressed and the solid solution phase is favoured over the intermetallic phases. This explanation and the name HEA came from Yeh and co-workers from year 2004 [10,11]. With proper composition designing, the HEA exhibits high hardness, excellent ductility as well as promising resistances to wear, oxidation and corrosion [5].

In this paper, we investigate VCrNiCoFeCu high-entropy alloy prepared by induction melting and spark plasma sintering methods. We conducted this comparative analysis for the two methods because the induction melting is not the best for industrial manufacturing due to the diseconomy and limitations in shape and size of final products. By contrast, mechanical alloying (MA) is a more convenient way, which has been widely used for the synthesis of nanocrystalline materials and widens the application scope of HEAs [6,7]. In addition, combined with the novel Spark Plasma Sintering (SPS) technique, bulk metallic glasses and high-entropy alloy can be easily obtained from the as-milled powders [8,9].

2. Experimental procedures

Metal powders of V, Cr, Ni, Co, Fe, Cu with purities of higher than 99.5 wt% and particle size ≤45 μm were used as initial materials. The elemental powders were mixed in equiatomic composition and milled in planetary ball miller (Fritsch pulverisette-4) for 20 h at 400 rpm in an argon atmosphere.

The obtained powder mixtures were sintered under vacuum using an SPS system (FCT System HPD 25). The sintering temperature was measured (0.05K precisely) and controlled by a thermocouple inserted into the graphite die wall at a distance of about 2 mm from the sintered sample. A uniaxial pressuring method was conducted using top and bottom graphite punches. The loading pressure was 90 MPa. A heating rate of 100 K/min was applied from room temperature to 480 °C before slowing down to 20 K/min from 480 °C to 500 °C to avoid temperature overshoot. The sintering temperature ranged from 500 to 550 °C. The holding time at the sintering temperature was in the range 20–30 min. Then the sample was removed from the heated region and a natural cooling was applied so that after about 30 min the specimen was at room temperature. The densified samples had a cylindrical shape, with a diameter of about 10 mm and a thickness of about 4 mm. All characterizations have been realized on these cylinders.

In the case of induction melting, lumps of high purity (99.99%) elements were melted under protecting atmosphere of argon to prepare the VCrNiCoFeCu equi-molar composition alloy.

The Scanning Electron Microscopy (SEM) was taken on FEI Quanta 3D FEG and Zeiss JSM microscopes equipped with Energy Dispersive Spectroscopy (EDS) and also Electron Back Scattered Diffraction (EBSD) instruments of EDAX.

The microstructure was investigated by electron backscatter diffraction (EBSD) using an FEI Quanta 3D scanning electron microscope (SEM). Before the EBSD investigation the surface was mechanically polished with 400, 600, 800, 1200, 2400 grit SiC abrasive. Finally, the surface was polished by 1 μm diamond paste

The chemical composition was examined by energy-dispersive X-ray spectroscopy (EDS) in SEM.

The structure of the alloy was identified by X-ray diffraction with Copper Kα (wavelength, λ=0.15418 nm) radiation in θ-2θ mode scanning from 20° to 80° at a scanning rate of 0.5°/min. For identification of the consistent phases, the PDF database was used. The average lattice parameter was determined by extrapolating the lattice parameters obtained from the different reflections to the diffraction angle of 2Θ= 180° using the Nelson–Riley method [12].

The corrosion resistances of the alloys were evaluated by polarization corrosion method. Potentiostatic polarization curves were measured with 10mV/min scan rate in 3.5% sodium chlorite solution (NaCl) at 25 °C in three-electrode electrochemical cell using a 20x5 mm polycrystalline Pt wire working electrode and an $Hg/Hg_2Cl_2/KCl$ reference electrode. The Tafel curve measurements were carried out using Zahner potentiostat. Electrochemical characteristic parameters, such as electrochemical corrosion potential (E_{corr}), corrosion current density (j_{corr}), can be obtained by evaluating polarization curves.

3. Results and discussion

3.1. Theory analysis and discussion

In this paper the enthalpy of mixing, atomic-size difference and the parameter Ω for the multi-component VCrNiCoFeCu alloy were calculated using the following equations [13-14]:

$$\Delta H_{mix} = \sum_{i=1, i\neq j}^{n} \Omega_{ij} c_i c_j \tag{1}$$

$$\Delta S_{mix} = -R\sum_{i=1}^{n} c_i Lnc_i \tag{2}$$

where $\Omega_{ij}(=4\Delta H^{mix}{}_{AB})$ is the regular melt-interaction parameter between *i*th and *j*th elements, c_i and c_j are the i-j atom concentrations, and $\Delta H^{mix}{}_{AB}$ is the mixing enthalpy of binary liquid alloys (the value can be obtained in [10-13]).

$$Delta(\delta) = \sqrt{\sum_{i=1}^{n} c_i (1 - r_i / \bar{r})^2} \tag{3}$$

where n is the number of the components in the alloy system, c_i is the atomic percentage of the *i*th component, $\bar{r}(=\sum_{i=1}^{n} c_i r_i)$ is the average atomic radius, and r_i is the atomic radius which could be obtained from [14-15].

The parameter Ω is defined to predict the solid-solution formation for various multi-component alloys as:

$$\Omega = \frac{T_m \Delta S_{mix}}{|\Delta H_{mix}|} \tag{4}$$

The melting temperature of n-elements alloy, T_m, is calculated using the rule of mixtures:

$$T_m = \sum_{i=1}^{n} c_i (T_m)_i \tag{5}$$

Here, $(T_m)_i$ is the melting point of the *ith* component of alloy. As analyzed above, the value of Ω is positive, and Ω=1 should be proposed as a critical value to form a solid solution. If Ω>1, the contribution of $T\Delta S_{mix}$ will exceed that of ΔH_{mix} for a solid-solution formation. If $\Omega\leq 1$, ΔH_{mix} is the dominant part in the equation, and formation of intermetallic compounds are expected [11].

In the case of VCrNiCoFeCu alloy this value is Ω=15.3.

Considering the small atomic size differences and that the value of Ω is larger than 1 indicates that the strain energy caused by lattice distortion is lower in VCrNiCoFeCu alloy and the entropy effect of ΔS_{mix} is large enough to exceed the ΔH_{mix}. Thus, the component atoms can randomly distribute in the alloy and a simple solid solution is stable [15].

Table 1. Calculated thermodynamic parameters of the VCrNiCoFeCu High Entropy Alloy.

Thermodynamic parameters	VCrNiCoFeCu
ρ (g/cm^3)	7.891
ΔH_{mix} (kJ/mol)	-1.7785
ΔG (J/mol)	-28976.76
ΔS_{mix} (J/K·mol)	14.90
T_m (K)	1825.6
Ω	15.30
δ (%) atomic radius	1.938
δ (%) WS radius	2.6654

In Table 1. **ρ** corresponds to density, $\mathbf{\Delta H_{mix}}$ is mixing enthalpy, **ΔG** the Gibbs free energy difference, $\mathbf{\Delta S_{mix}}$ the mixing enthalpy, $\mathbf{T_m}$ is the melting point, **Ω** is the parameter for solid solution formation and **δ** is the atomic size difference calculated in both covalent atomic radii and WS cell radius parameters. Calculating the ΔH_{mix} and δ parameters for a numerous HEA's and metallic glasses, Guo and Liu [14] (see Fig.1 in [14]) could delimitate the following regions in ΔH_{mix} versus δ diagram: HEA with a solid solution phase form when the three parameters are in the range of: $-22 \leq \Delta H_{mix} \leq 7$ kJ/mol, and $\delta < 8.5$ and $\Delta S_{mix} \leq 19.5$ J/(K·mol).

In the case of VCrNiCoFeCu alloy the atomic-size difference δ = 1.938 (in covalent atomic radius) the mixing enthalpy ΔH_{mix}= -1.7785 kJ/mol and ΔS_{mix} = 14.90 J/ (K·mol), respectively, as can be seen Table 1.

The values of Ω and δ for VCrNiCoFeCu match the criteria to form a solid solution.

3.2. Electrochemical properties and variation in the microstructure after the test

The open circuit potential (OCP) and polarization curves with electrochemical characteristic parameters of the investigated alloy in 3.5% sodium chlorite (NaCl) solution are shown in Fig. 1 and 2. The corrosion rates and electrochemical parameters of alloys in 3.5% NaCl solution are presented in Table 2 and 3. The corrosion potentials for induction melted sample are varying from -0.5489 to -0.6539 V_{SCE} and for the SPSed sample from -0.5846 to 0.6921 $V_{SCE,}$ as can be seen in Tables 2 and3.

Corrosion rate obtained from the electrochemical polarization measurements can be calculated by:

$$Corrosion \text{ rate (mm/y)} = \frac{3.27 \times 10^{-3} \times j_{corr} \times EW}{\rho} \quad (6)$$

where j_{corr} (in μA/cm^2) is the corrosion current density, EW the equivalent mass of the alloy, and ρ is the density of the alloy. EW is defined as follows [16],

$$EW = \frac{1}{\sum \left(\frac{f_i n_i}{a_i} \right)} \quad (7)$$

where f_i, n_i, and a_i are mass fraction, electrons exchanged, and atomic mass, respectively, of the *i*th alloying element. The EW value of VCrNiCoFeCu is 31.157 and the corrosion rate (mm/y) as a function of the time can be found in Tables 2. and 3.

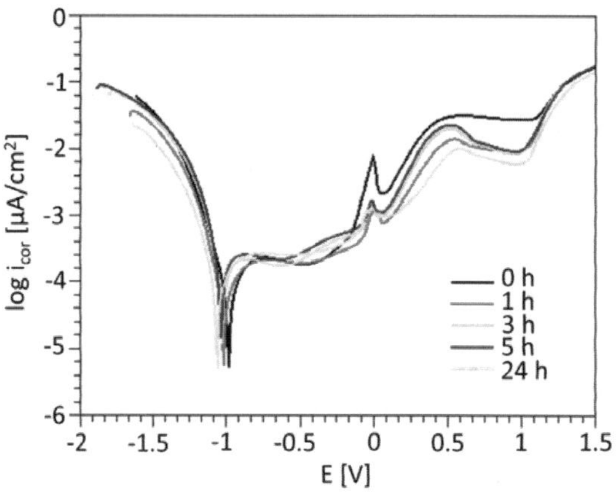

Figure 1. Polarization curves for induction melted VCrNiCoFeCu alloy in 3.5% NaCl solution

Table 2. Electrochemical parameters of the VCrNiCoFeCu alloy, prepared by induction melting method in 3.5% NaCl solution at room temperature (25°C).

Time	E_{corr} (V)	j_{cor} (mA/cm^2)	CR (mm/year)	A (cm^2)
0h	-0.6539	3.2866	0.042044244	0.2
1h	-0.5776	3.5541	0.045466272	
3h	-0.5871	3.2484	0.041555566	
5h	-0.5489	3.2325	0.041352163	
24h	-0.5585	3.3822	0.043267219	

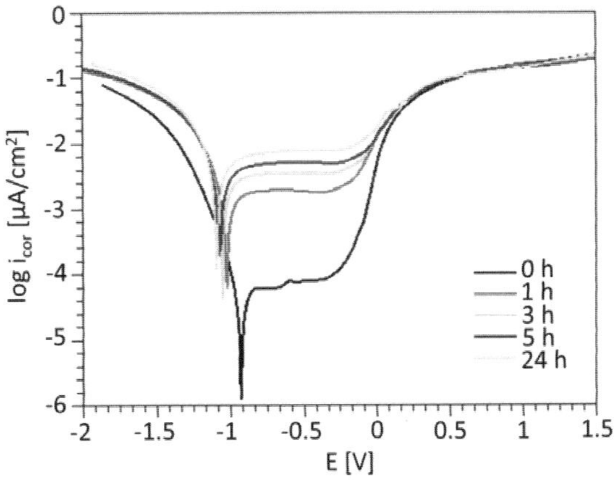

Figure 2. Polarization curves for SPS VCrNiCoFeCu alloy in 3.5% NaCl solution

Table 3. Electrochemical parameters of the VCrNiCoFeCu alloy, prepared by SPS method in 3.5% NaCl solution at room temperature (25°C).

Time	E_{corr} (V)	j_{cor} (mA/cm^2)	CR (mm/year)	A (cm^2)
0h	-0.5846	3.435	0.04394267	0.35
1h	-0.6443	2.5612	0.032764473	
3h	-0.6524	2.2113	0.028288334	
5h	-0.6621	2.1964	0.028097723	
24h	-0.6921	2.1051	0.026929757	

For both alloys three specimens were tested to check the repeatability of the test results. Lower corrosion rate and higher breakdown voltage can be seen in the case of SPSed sample. The transient effect continues until the surface film thins to a degree that it becomes stable with respect to the environment. In the case of induction melted sample corrosion rates were almost stable during 24 h measurements. The differences in chemistry between these two alloys may influence the corrosion performance and this difference is regarded to the preparation methods.
The achieved corrosion resistance of both HEA alloys can be regarded to a protective layer of chromium oxide that has evolved on the surface. In the case of induction-melted sample, this film so thin that it is even not visible by XRD, on the other hand the SEM-EDX demonstrated its presence, as can be seen Figures 5-6.
The pitting corrosion resistance may be quantified using the passive breakdown potential, since sustained breakdown of the passive film is indicative of the onset of pitting corrosion.

This can be seen after the passive regions from which current decreases rapidly with increasing potential. The significantly higher breakdown potential for this HEA designates greater protection against localized corrosion in the chloride-containing solutions, than that of the SS304 [8]. The spikes in current that are seen in Fig. 1. and Fig. 2 may be attributed to metastable pitting. These current fluctuations are caused by corrosion pits, which re-passivize ultimately.

3.3. Microstructure and phase constitution of VCrNiCoFeCu alloys
The XRD results for both alloys are shown on Figures 3-4. The results suggest that both the induction melted and SPS made HEAs are single-phase FCC alloys. It should be mentioned that the in the case of SPS method the sample was polluted by WC (tungsten carbide), which came from the milling jar. The XRD profile of SPS sample in Fig. 4. show significant phase of WC.
The average lattice parameter was determined from the diffraction peak positions using the Nelson-Riley method [18]. The measured X-ray diffractograms indicated both samples have single-phase face-centered cubic (FCC) structure with lattice parameters of a= 0.3612 ± 0.0002 nm and a=0.3665 ± 0.0002, for induction melted and SPSed samples, respectively.

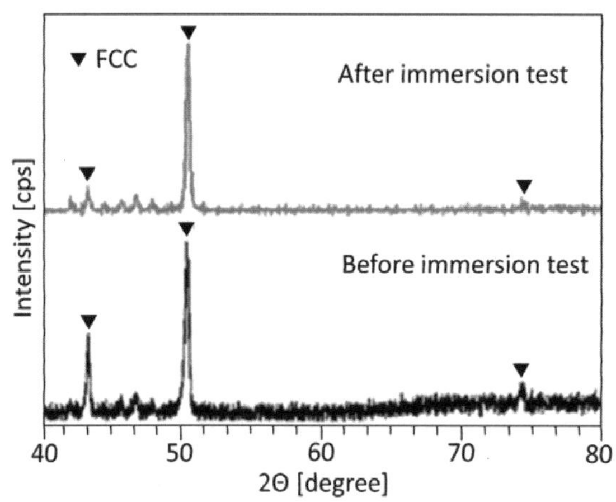

Figure 3. X-ray diffraction patterns of the induction melted VCrNiCoFeCu HEA

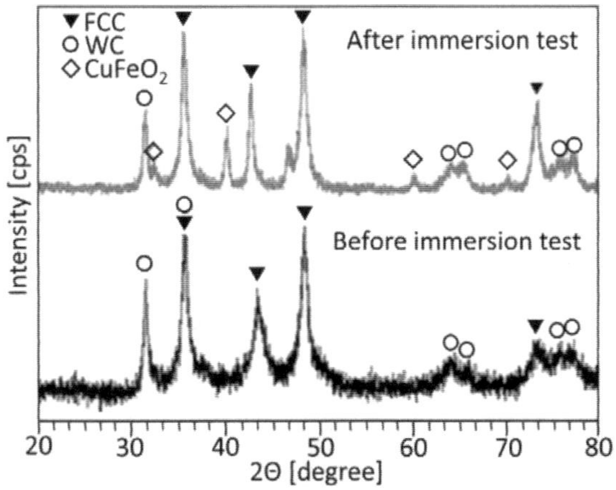

Figure 4. X-ray diffraction patterns of the SPS VCrNiCoFeCu HEA

The SEM micrographs of these alloys are shown in Figures 5-6. Microstructure also plays an important role in corrosion behaviour. Compositional differences which became from the mechanical alloying (the SPS sample was polluted by WC) may result in galvanic coupling, which can be much important than the effect of microstructure on general corrosion.

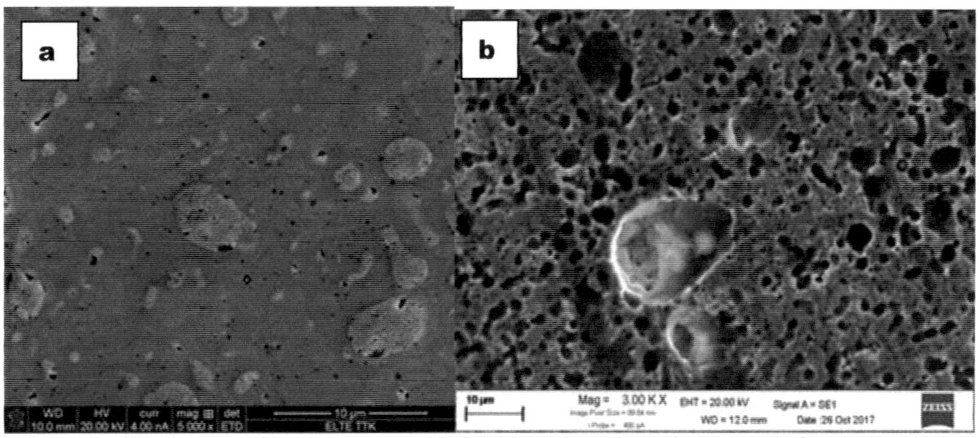

Figure 5. SEM picture of induction melted VCrNiCoFeCu HEA before (a) and after (b) immersion test in 3.5%NaCl solution.

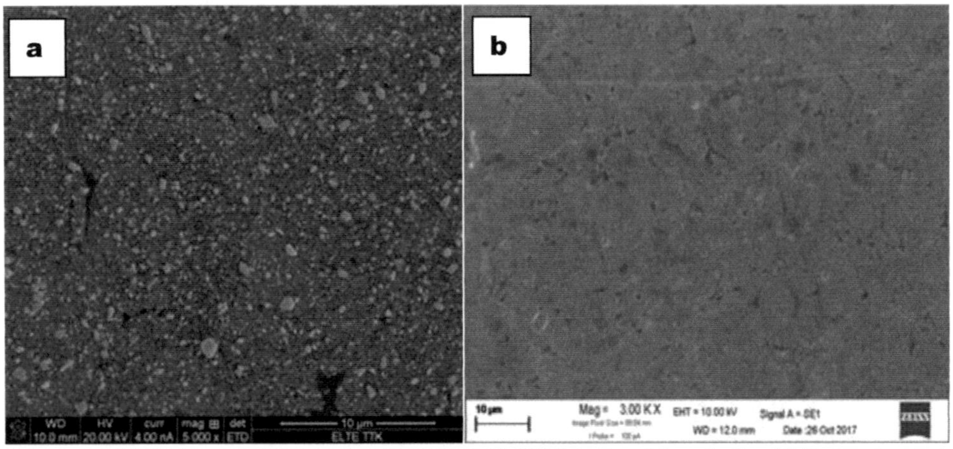

Figure 6. SEM picture of SPS made VCrNiCoFeCu HEA before (a) and after (b) immersion test in 3.5%NaCl solution.

On the SEM figure of induction melted sample (Figure 5.) some chemical inhomogeneity can be seen before the corrosion test. After immersion a thin, high Cr containing film was formed as a protective coating for this alloy. On the Figure 6 one can see the SEM pictures for SPSed sample. The picture of ETD detector shows the surface morphology by secondary electrons. The brighter ones on the picture can be regarded to tungsten carbide pollution (from the milling jar). After the immersion these artefacts are removed, only a layer with high Cr content can be seen with some lower scale pitting corrosion.

4. Conclusions

The two samples of VCrNiCoFeCu prepared by induction melting and spark plasma sintering are crystallized in FCC structure. In the case of induction melted sample, high porosity can be observed. The phenomena can be regarded to Cu segregation effect, and to the fact, that due to their higher viscosity, HEAs tend to have higher porosity than conventional alloys (like steels).
As it can be seen in the Figure 5, a pitting corrosion is observable on the surface after immersion test in 3.5% NaCl solution (for 24 hours). This pitting corrosion behaviour can be attributed to galvanic action between dendrite and inter-dendrite compositions.
In the case of spark plasma sintered alloy, the dendrite like structure is not observable. This can be due to high WC contamination, what is inevitable in the case of mechanical milling.

5. Acknowledgements

This work was supported in part by EIT KIC Raw Materials NoI 15053 "EXTREME" project coordinated by ENEA.
The work was carried out with financial support in part from the Russian Science Foundation, project No. 17-73-20272.

6. References

[1] D V Louzguine-Luzgin, L V Louzguina-Luzgina, T Saito, G Xie, A Inoue 2008 *Structure and properties of high strength and ductile Ti-Fe-Cu-Nb-Sn alloys* Mater. Sci. Eng. A. vol 497 p 126–131 doi:10.1016/j.msea.2008.06.020
[2] B B Sun, M L Sui, Y M Wang, G He, J Eckert, E Ma 2006 *Ultrafine composite microstructure in a bulk Ti alloy for high strength, strain hardening and tensile ductility* Acta Mater vol 54 p 1349–1357 doi:10.1016/j.actamat.2005.11.011

[3] Y Lu, Y Dong, S Guo, L Jiang, H Kang, T Wang, B Wen, Z Wang, J Jie, Z Cao, H Ruan, T Li 2014 *A promising new class of high-temperature alloys: eutectic high-entropy alloys* Sci. Rep vol 4 p 6200 doi:10.1038/srep06200

[4] S Y Chang, M K Chen 2009 *High thermal stability of AlCrTaTiZr nitride film as diffusion barrier for copper metallization* Thin Solid Films vol 517 p 4961–4965 doi:10.1016/j.tsf.2009.03.078

[5] M H Tsai, J W Yeh, J Y Gan 2008 *Diffusion barrier properties of AlMoNbSiTaTiVZr high-entropy alloy layer between copper and silicon* Thin Solid Films vol 516 p 5527–5530 doi:10.1016/j.tsf.2007.07.109

[6] C Ng, S Guo, J Luan, S Shi, C T Liu 2012 *Entropy-driven phase stability and slow diffusion kinetics in an Al 0.5CoCrCuFeNi high entropy alloy* Intermetallics vol 31 p 165–172 doi:10.1016/j.intermet.2012.07.001

[7] É Fazakas, J Q Wang, V Zadorozhnyy, D V Louzguine-Luzgin, L K Varga 2014 *Microstructural evolution and corrosion behavior of* $Al_{25}Ti_{25}Ga_{25}Be_{25}$ *equi-molar composition alloy* Mater Corros vol 65 doi:10.1002/maco.201206941

[8] Y J Hsu, W C Chiang, J K Wu 2005 *Corrosion behavior of FeCoNiCrCux high-entropy alloys in 3.5% sodium chloride solution* Mater Chem Phys vol 92 p 112–117 doi:10.1016/j.matchemphys.2005.01.001

[9] Y Y Chen, T Duval, U D Hung, J W Yeh, H C Shih 2005 *Microstructure and electrochemical properties of high entropy alloys-a comparison with type-304 stainless steel* Corros. Sci. vol 47 p 2257–2279 doi:10.1016/j.corsci.2004.11.008

[10] P. Huang, J. Yeh 2004 *Multi-Principal Element Alloys with Improved Oxidation and Wear* Resistance for Thermal Spray Coating, Adv. Eng. Mater. 6 74–78. doi:10.1002/adem.200300507

[11] J W Yeh, S K Chen, S J Lin, J Y Gan, T S Chin, T T Shun, C H Tsau, S Y Chang 2004 *Nanostructured high-entropy alloys with multiple principal elements: Novel alloy design concepts and outcomes* Adv. Eng. Mater. vol 6 p 299–303 doi:10.1002/adem.200300567

[12] J Nelson and D Riley 1945 *An experimental investigation of extrapolation methods in the derivation of accurate unit-cell dimensions of crystals* Proc Phys Soc Lond vol 57 p 160 doi: 10.1088/0959-5309/57/3/302

[13] K Y Tsai, M H Tsai, J W Yeh 2013 *Sluggish diffusion in Co-Cr-Fe-Mn-Ni high-entropy alloys Acta Mater vol 61 p 4887–4897 doi:10.1016/j.actamat.2013.04.058*

[14] D B Miracle, O N Senkov 2017 *A critical review of high entropy alloys and related concepts* Acta Mater vol 122 p 448–511 doi:10.1016/j.actamat.2016.08.081

[15] H Oh, D Ma, G Leyson, B Grabowski, E Park, F Körmann, D Raabe 2016 *Lattice Distortions in the FeCoNiCrMn High Entropy Alloy Studied by Theory and Experiment* Entropy vol 18(9) p 321 doi:10.3390/e18090321

[16] Y J Zhou, Y Zhang, F J Wang, G L Chen 2008 *Phase transformation induced by lattice distortion in multiprincipal component CoCrFeNiCuxAl1-x solid-solution alloys* Appl Phys Lett vol 92 241917. doi:10.1063/1.2938690.

[17] F J Wang, Y Zhang, G L Chen 2009 *Atomic packing efficiency and phase transition in a high entropy alloy* J Alloys Compd vol 478 p 321–324 doi:10.1016/j.jallcom.2008.11.059

[18] J Nelson, D Riley 1945 *An experimental investigation of extrapolation methods in the derivation of accurate unit-cell dimensions of crystals* Proc. Phys. Soc. Lond. vol 57 p 160 doi: 10.1088/0959-5309/57/3/302

IOP Conf. Series: Materials Science and Engineering **329** (2018) 012017 doi:10.1088/1757-899X/329/1/012017

Wear Resistance Increase by Friction Stir Processing for Partial Magnesium Replacement in Aluminium Alloys

Sebastian Balos*, Danka Labus Zlatanovic, Petar Janjatovic, Miroslav Dramicanin, Dragan Rajnovic, Leposava Sidjanin

Faculty of Technical Sciences, University of Novi Sad, Trg Dositeja Obradovica 6, 21000 Novi Sad, Serbia

Corresponding author: sebab@uns.ac.rs

Abstract. In this paper, the influence of friction stir processing (FSP) was evaluated as a way of increasing mechanical properties and a way of replacing the magnesium content in aluminium alloys. FSP was done on AA5754 H111 aluminium alloy, containing 3 % Mg, by using various types of tools and different welding speeds, rotational speeds and tilt angles. Wear test was done against SiC abrasive papers. SiC was used to simulate extreme abrasive wear conditions. The wear test was done on untreated AA5754 specimens, processed AA5754 specimens and untreated AA5083 H111 specimens, the latter containing 4.5 % Mg. AA5083 was chosen as an alternative to AA5754, but with a significantly higher Mg content. Base material microhardness was 60 HV1 and 80 HV1 for AA5754 and AA5083 alloys respectively. To find the effect of FSP on AA5754 alloy, microstructures were studied, mainly grain size in the stir zone. It was found, that an elevated processing and rotational speed, without tilt angle and the tool without a reservoir resulted in an increase in hardness of the AA5754 to 70 HV1, but with the occurrence of tunneling defect and the wear rate of 79.3 mg. Lower FSP parameters and a tilted tool with a reservoir resulted in microhardness of 68 HV1 and wear rate of 68.2 mg without tunneling. These wear values are lower than those obtained with unmodified Al-alloys: AA5754 97.2 mg and AA5083 86.3 mg. An increased wear resistance can be attributed to the combined effect of grain boundary strengthening mechanism and solid solution strengthening, versus only the latter in untreated alloys.

1. Introduction

The relatively high specific strength and particularly specific stiffness of aluminium and its alloys due to low density make them suitable for a significant number of applications in aerospace and automotive industries. There has been a tendency to improve these properties even more, because of increasing need of high strength to weight ratio. However, in increase in demand for aluminium alloys is in contrast to the most common alloying elements availability. Namely, one of the most common alloying elements is magnesium, which is considered to be a critical raw material (CRM) for European Union [1]. One of the accepted methods for improving mechanical properties alloys is grain refinement strengthening mechanism through friction stir processing (FSP). It has several advantages: it can be used to fabricate metal matrix composites; microstructure and mechanical properties of the processed zone can be accurately controlled by optimizing the tool design and process parameters, providing good mechanical properties in the as-processed condition [2]. It can be used for superplastic forming (SPF) by achieving the high-strain rate superplasticity (HSRS) at lower temperatures [3]. Also, it can be easily automated on specialized or simple universal milling machines, with a relatively low environmental impact [4, 5].

The FSP is a variation of friction stir welding (FSW), a solid state joining process invented at The Welding institute (TWI), UK in 1991, as a technique for joining Al alloys. However, instead of FSW the FSP was developed to modify the microstructure of metalic parts. [6-11]. The FSP has been successfully applied to modify the microstructure of various cast metal, to improve the mechanical properties [11-15]. For example, in [12] after FSP, yield and tensile strength were improved by thirty percent over the base aluminium 2285 alloy. In casts A319 and A356, FSP was applied to reduce porosity and to create more uniform distributions of second-phase particles. As result, the ultimate tensile strengths, ductilities, and fatigue lives of both alloys were increased [13]. In [14], the hardness and tensile strength of the friction stir-processed 1050 aluminum alloy increased significantly. However, since the technique is relatively new there is certainly a need for a better scientific understanding.

The aim of this paper is to study the influence of different tool geometries and process parameters on mechanical properties and microstructure of Al-Mg alloy. The main aim is to develop fine- grained microstructure by the FSP, which reflects an increase in hardness and wear resistance, compared to a Al-Mg alloy with a higher Mg content in order to provide a reduction of Mg which is a CRM.

2. Experimental

In this study, the specimens used for the friction stir processing were made of EN AW 5754 H111 plates, 8 mm thick, having the dimensions 130 mm × 100 mm. The chemical composition of the aluminium alloy was determined by using ARL 3580 optical emission spectrometer, Table 1. In Table 1, the chemical composition of the comparative EN AW 5083 H111 is shown. The specimens were placed into a steel fixture within a 130 mm groove and secured in place by several clamps. The fixture was fitted onto an adapted Prvomajska UHG universal milling machine. The tools used were made of H11 hot-work tool steel, with the hardness of 62±2 HRC. Tool geometry is shown in Figure 1. Three types of tool geometries were used: a) with a flat shoulder and cylindrical pin with a thread, b) with concave shoulder and square pin and c) with three concentric reservoirs with a square pin. The FSP parameters are shown in Table 2. The plunge depth of tool shoulder was 0.2 mm for all FSP samples.

The properties of the FSP samples were determined using microhardness testing, metallographic examination and wear testing. Microhardness was determined with Wilson Tukon 1102, with a 1-kg load. To obtain the microhardness profiles the microhardness measurements were done using a 0.5 mm distance between the indentations and 1 mm from the surface as shown on Figure 2. The metallographic examination was done after the standard preparation procedure: grinding with sandpapers (grit 220 to 2000), polishing with diamond suspensions (6, 3, 1 and 1/4 μm abrasive-grain sizes) and etching with Keller's reagent (2 mL HF, 3 mL HCl, 5 mL HNO3, 190 mL H_2O). The metallographic samples were examined with a Leitz Orthoplan light microscope. The wear testing was conducted on a pin-on-disc method, with the following parameters: sample size 10×10 mm; SiC grinding paper with grit size P600 for 1 min under 1kg load. The specimen followed the path 70 mm away from the abrasive paper center, with water flow maintained at 10 ml/min. Testing specimens followed the same wear path. Spindle speed was 250 min^{-1}, providing 110 mm wear path. Before testing, each specimen was ground up to 2000 grit abrasive paper, to the surface Ra of 0.07 μm. Each wear test was conducted on three specimens and the average mass loss was reported.

Table 1. Chemical composition of aluminium alloys (in weight %)

Alloys	Cu	Mn	Mg	Si	Fe	Zn	Ti	Cr
EN AW 5754	0.05	0.33	3.10	0.33	0.37	0.06	0.025	0.039
EN AW 5083	0.07	0.57	4.10	0.26	0.32	0.07	0.026	0.10

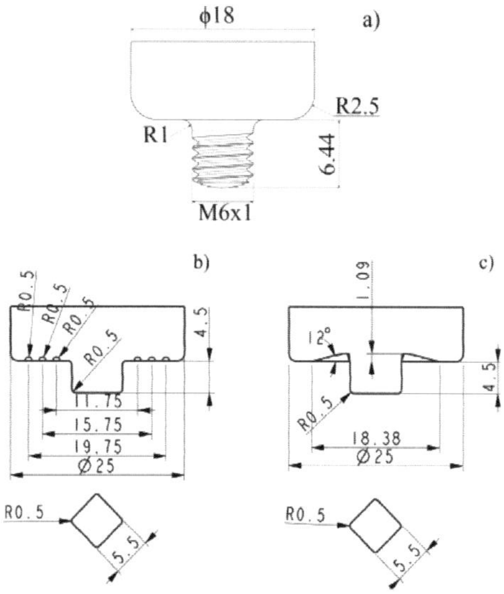

Figure 1. FSP tools: a) cylindrical profile with a flat shoulder with tread (type I), b) feature shoulder with three concentric circles with the overall-volume-to-pin ratio of 0.5 (II-type) and c) concave shoulder with the shoulder-to-pin volume ratio of 0.5 (III-type).

Table 2. Friction stir processing parameters

Specimen designation	**Rotational speed [min^{-1}]**	**Processing speed [mm/min]**	**Tool tilt [°]**	**Tool type**
1	710	40	0	I
2	710	20	0	I
3	560	40	0	I
4	560	20	0	I
5	560	40	3	I
6	560	20	3	I
7	450	40	3	I
8	450	20	3	I
9	450	20	3	II
10	450	20	3	III

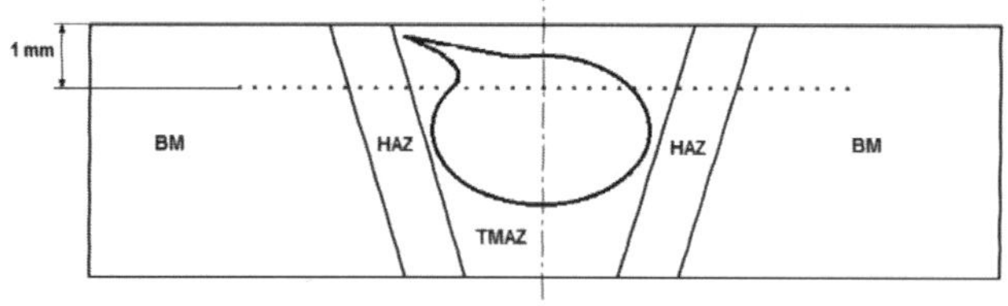

Figure 2. Hardness measurement distribution scheme

3. Results and discussion

3.1 Macro imagery

The metallographic images of samples processed with tool I and the processing speed of 40 mm/min are presented in Figure 3. In Figure 3 a,b samples without tilt angle and with rotational speed of 710 min^{-1} and 560 min^{-1} are depicted. Samples with tilt angle 3° and with rotational speeds 560 min^{-1} and 450 min^{-1} are shown in Figure 3 c,d. It can be seen that the tunnel-like defects occur in all samples in Figure 3. The defects become smaller as the rotational speed is reduced and the tilt angle is introduced. These defects are typical for FSW/P samples if the processing conditions i.e. processing speed, tool rotation etc. fail to generate the required heat for bonding, inadequate material mixing and stirring can occur, resulting in the formation of tunnel defects [16, 17]. In [17], it has been shown that most notable influence on the size and shape of the defects have rotational and processing speed and plunge downforce. A similar trend can be seen in Figure 4, where macro images of samples processed tool type I and 20 mm/min processing speed. In Figure 4 a,b samples without tilt angle and with rotational speeds 710 min^{-1} and 560 min^{-1}are given while in Figure 3c,d, samples with tilt angle of 3° and with rotational speeds 560 min^{-1} and 450 min^{-1}are shown, respectively.

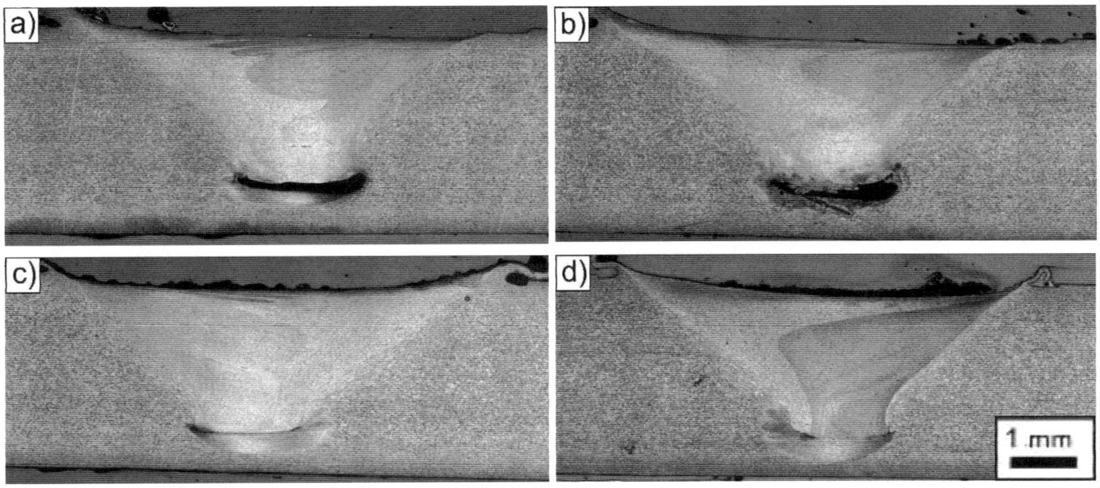

Figure 3. Macrographs of samples processed with tool type I; processing speed 40 mm/min and different rotational speeds (ω): a) with ω=710 min^{-1} and without tilt angle; b) with ω=560 min^{-1} and without tilt angle; c) with ω=560 min^{-1} and with tilt angle 3°; d) with ω=450 min^{-1} and with tilt angle 3°.

In all samples tunnel-like defects occur in bottom section of the NZ (nugget zone). The macrostructures of the HAZ (heat affected zone) and TMAZ (thermomechanical zone) zones are similar in all samples processed with tool type I. Also in all samples the NZ has finer grain structure compared to basic material.

In Figure 5, samples processed with 20 mm/min, 450 min^{-1}, tilt angle 3° and tool type II (Figure 5a) and type III (Figure 5b) are shown. In both samples, no tunnel-like defects are present. Microstructure in the NZ is also finer compared to the basic material.

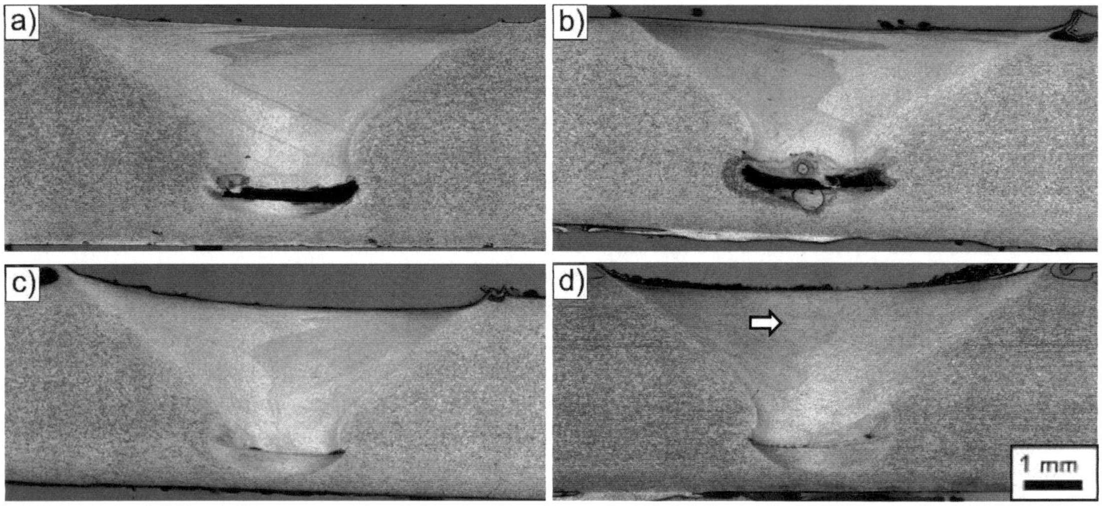

Figure 4. Macrographs of samples processed with tool type I; processing speed 20 mm/min and different rotational speeds: a) ω=710 min^{-1} and without tilt angle; b) with ω=560 min^{-1} and without tilt angle; c) with ω= 560 min^{-1} and with tilt angle 3°; d) with ω=450 min^{-1} and with tilt angle 3°.

In Figure 6. it can be seen that the microstructure in the NZ in samples processed with tool type I is finer than that of tool types II and III. Furthermore, the microstructure of sample processed with tool type II is slightly finer than microstructure obtained with tool type III.

All samples processed with the tool with a flat shoulder Figure 1a developed a tunnel-like defects. This phenomenon is the result of the tool-shoulder geometry and, therefore, its influence on the material flow. Even a relatively small reservoir Figure 1c and d provides a more convenient material flow. This allows the material to move not only perpendicularly to the tool axis, as forced by the pin, but also in parallel to the tool axis, eliminating the tunnel-like defect at a lower processing and rotational speed [9, 11, 15]. At a constant rotational speed, a relatively low processing speed causes an increase in the stirring-impulse frequency at a given processed length, leading to a more effective filling and defect avoidance. These results support the findings from [18], where the influence of a tunneling-type defect on the mechanical and metallurgical properties of an Al-Mg alloy friction stir welded was studied.

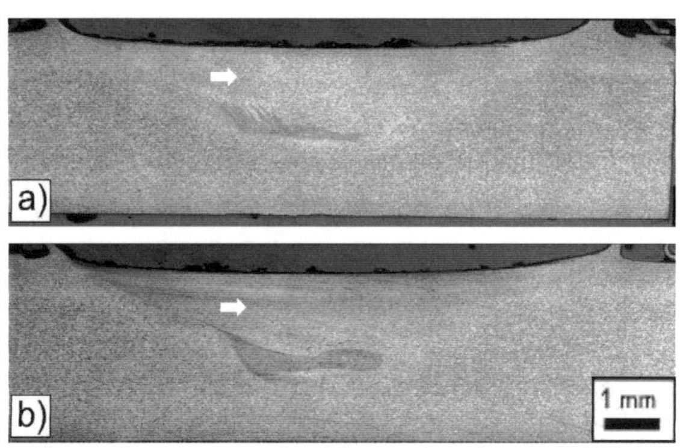

Figure 5. Macrographs of samples processed with rotational speed 450 min^{-1}; processing speed 20 mm/min; tilt angle 3°: a) tool type II; b) tool type III.

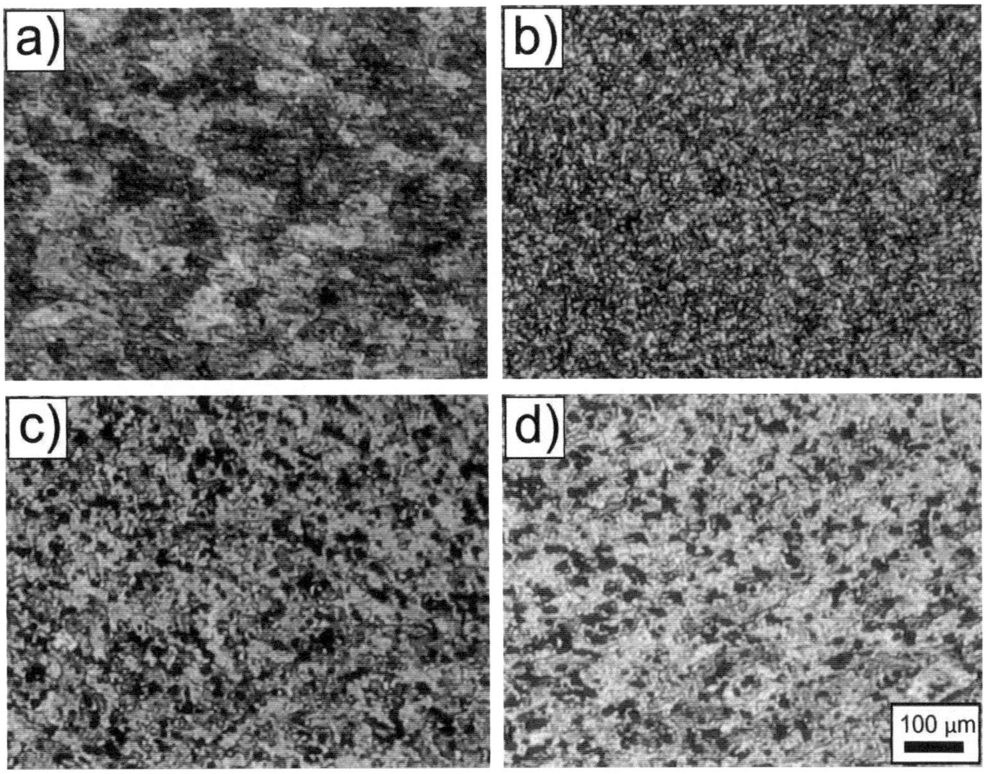

Figure 6. Microstructure of processed samples at locations shown in Figures 4 and 5 indicated by white arrows at processing speed 20 mm/min, rotational speed 450 min^{-1}: a) base material; b) tool type I; c) tool type II d) tool type III

3.2 Hardness profiles

Hardness profiles of FSP specimens are presented in Figures 7 and 8. Hardness profiles of samples processed with tool type I are shown in Figure 7. Hardness profiles have a similar shape, with the maximum attained hardness values at the middle of the chart, that is, in the NZ. These results are supported by the microstructural analysis which indicates that in the stir zone (NZ), the highest refinement was observed. Also, the hardness in the NZ is higher for the samples processed at smaller processing speeds. Finally, the decrease in the rotational speed and the introduction of tilt angle causes a drop in hardness.

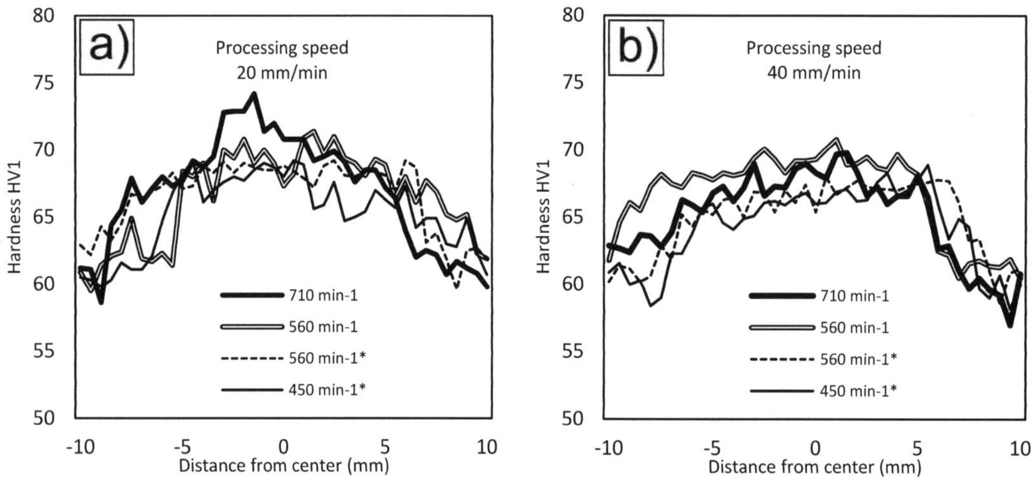

Figure 7. Microhardness profile of samples processed with tool type I; processing speed: a) 20 mm/min and b) 40 mm/min and different rotational speeds, tilt angle 3°

Hardness profile of samples processed with tool types II and III are depicted in Figure 8. There are generally slightly lower hardness values in sample processed with tool geometry II. The hardness values in Figure 8. are lower than profiles in Figure 7 for the same processing and rotational speeds. Most notably, the area with increased hardness is wider, due to a wider shoulder in type I tool. This, together with a shallower pin, with a longer travel of material in vertical plane, results in a more uniform hardness distribution in tool types II and III.

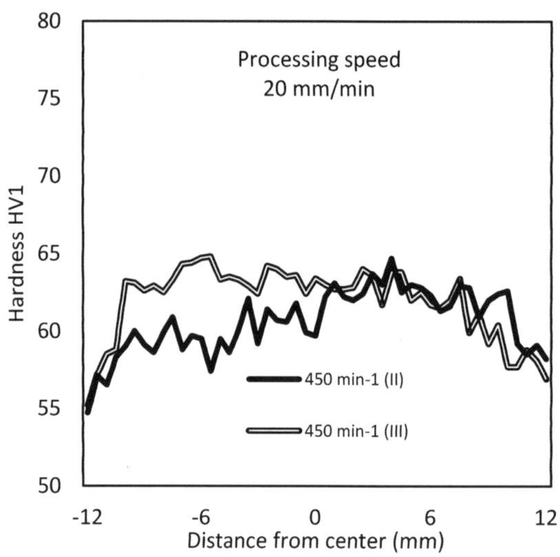

Figure 8. Microhardness profile of samples processed with tool type II and III; processing speed 20 mm/min, rotational speed 450 min^{-1} and with tilt angle 3°

3.3 Wear testing

Wear testing results are shown in Table 3. Wear testing was done on the base metal (EN AW 5754), the comparative material with increased Mg content (EN AW 5083). Also, wear rate was measured on two of the processed specimens. Firstly, the specimen 10, without tunneling defect, obtained with tool type III, as the one having a higher hardness in the majority of areas compared to the specimen obtained with tool type II. Finally, the wear rate was determined on specimen 2, processed with tool type I, rotation speed 710 mm^{-1} and processing speed of 20 mm/min. This particular specimen is of interest, since the tunneling defect is the most pronounced. Tunneling of this extent could be of interest for compact heat exchangers [19].
Wear rates in terms of mass loss of selected specimens is shown in Table 3. It can be seen that the highest wear rate, that is, the lowest wear resistance has the EN AW 5754 base metal, followed by the harder EN AW 5083 alloy. Friction stir processed specimens 2 and 10 performed the best, although their hardness does not exceed the hardness of the EN AW 5083 alloy. The lowest mass loss was measured in specimen 10, obtained with polygonal tool with reservoir without the tunneling. This relatively high wear resistance compared to unprocessed specimens can be the result of the combined work hardening, grain boundary strengthening mechanism originated from the friction stir process and solid solution strengthening from the alloy itself, compared only to solid solution strengthening in the untreated alloys.

Table 3. Mass loss of wear tested specimens

	EN AW 5754			**EN AW 5083**
	Base material; 60 HV1	Specimen 2; 70 HV1	Specimen 10; 68 HV1	Base material; 80 HV1
Mass loss [mg]	97.2±5 mg	79.3±6 mg	68.2±5 mg	86.3±6 mg

4. Conclusions

According to the presented results and within the limitations of the experiment setup, the following conclusions can be drawn:

- An elevated processing and rotational speed, without tilt angle and the tool without a reservoir resulted in an increase in hardness of the AA5754, but with the occurrence of tunneling defect and wear rate of 79.3 mg.
- Lower FSP parameters and a tilted tool with a reservoir resulted in microhardness of 68 HV1 and wear rate of 68.2mg without tunneling.
- These wear values are lower than those obtained with unmodified Al-alloys: AA5754 97.2 mg and AA5083 86.3mg.
- An increased wear resistance can be attributed to the combined effect of grain boundary strengthening mechanism and solid solution strengthening, versus only the latter in untreated alloys.

References

[1] Grilli ML, Bellezze T, Gamsjäger E, Rinaldi A, Novak P, Balos S, Piticescu RR and Ruello ML 2017 *Materials* **10** 285-307
[2] Fujji H, Cui Ling, Maeda M, Nogi K 2006 *Mater. Sci. Eng. A* **419** 25–31
[3] Smoliej A, Kločar D, Skaza B, Nagode A, Salček E, Dragojevic V, Smolej S 2014 *Mater. Sci. Eng. A* **590** 239–45
[4] Ma ZY, Pilchak AL, Juhas MC and Williams JC 2008 *Scr. Mater.* **58** 361–66
[5] Yadav D and Bauri R 2012 *Mater. Sci.Eng. A*, **539** 85–92
[6] Mishra RS, De PS, Kumar N 2014- *Friction Stir Welding and Processing book*
[7] Mishra RS and Ma ZY 2005 *Mat. Sci. Eng. R* **50** 1-7

[8] Ma ZY 2008 *Metall Mater. Trans. A* **39a** 642-58
[9] Balos S, Sidjanin L, Dramicanin M, Labus Zlatanovic D, and Antic A 2016 *Mater. Tehnol.* **50** 387-94
[10] Ma ZY, Sharma SR and Mishra RS 2006 *Mater. Sci. Eng. A* **433** 269–78
[11] McNelley TR, Swaminathan S, and. Su JQ 2008 Scr. Mater. 58 349-54
[12] Karthikeyan L, Senthilkumar VS, Balasubramanian V and Natarajan S 2009 *Mater. Des.* **30** 2237–42
[13] Santella ML, Engstrom T, Storjohann D, Pan TY 2005 *Scr. Mater.* **53** 201–6
[14] Kwon YJ, Shigematsu I. and Saito N 2003 *Scr. Mater.* **49** 785–9
[15] Ramesh Babua S, Pavithranb S, Nithinb M and Parameshwaranb B 2014 12th *Global Congress on Manufacturing and Management* (*Procedia Engineering 97*) p 800–9
[16] Kah P, Rajan R, Martikainen J, Suoranta R 2015 *Int. J. Mech Mater. Eng.* DOI 10.1186/s40712-015-0053-8
[17] Kim YG, Fujji H, Tsumura T, Komayaki T, Nakata K 2006 *Mat. Sci. Eng. R* **415** 250-4
[18] Balos S and Sidjanin L 2014 *Mater. Tehnol.* **48** 491–6
[19] Mehendale, SS, Jacobi, AM, Shah, RK, 2000 *App. Mech. Rev.* **53** 175-93